Gels Horizons: From Science to Smart Materials

Series Editor

Vijay Kumar Thakur, School of Aerospace, Transport and Manufacturing, Cranfield University, Cranfield, Bedfordshire, UK

This series aims at providing a comprehensive collection of works on the recent advances and developments in the domain of *Gels*, particularly as applied to the various research fields of sciences and engineering disciplines. It covers a broad range of topics related to *Gels* ranging from *Polymer Gels, Protein Gels, Self-Healing Gels, Colloidal Gels, Composites/Nanocomposites Gels, Organogels, Aerogels, Metallogels & Hydrogels* to *Micro/Nano gels*. The series provides timely and detailed information on advanced synthesis methods, characterization and their application in a broad range of interrelated fields such as chemistry, physics, polymer science & engineering, biomedical & biochemical engineering, chemical engineering, molecular biology, mechanical engineering and materials science & engineering.

This Series accepts both edited and authored works, including textbooks, monographs, reference works, and professional books. The books in this series will provide a deep insight into the state-of-art of *Gels* and serve researchers and professionals, practitioners, and students alike.

More information about this series at http://www.springer.com/series/15205

Amit Kumar Nayak · Md Saquib Hasnain ·
Dilipkumar Pal
Editors

Ionically Gelled Biopolysaccharide Based Systems in Drug Delivery

Springer

Editors
Amit Kumar Nayak
Department of Pharmaceutics
Seemanta Institute of Pharmaceutical Sciences
Jharpokharia, Odisha, India

Md Saquib Hasnain
Department of Pharmacy
Palamau Institute of Pharmacy
Daltonganj, Jharkhand, India

Dilipkumar Pal
Department of Pharmaceutical Sciences
Guru Ghasidas Vishwavidyalaya
Bilaspur, Chhattisgarh, India

ISSN 2367-0061 ISSN 2367-007X (electronic)
Gels Horizons: From Science to Smart Materials
ISBN 978-981-16-2270-0 ISBN 978-981-16-2271-7 (eBook)
https://doi.org/10.1007/978-981-16-2271-7

This Springer imprint is published by the registered company Springer Nature Singapore Pte Ltd.
The registered company address is: 152 Beach Road, #21-01/04 Gateway East, Singapore 189721, Singapore

Preface

In recent years, ionic gelation of ionic biopolysaccharides (such as alginate, pectin, gellan gum and chitosan) has been exploited to develop various drug-releasing carriers. Ionic gelation of ionic biopolysaccharides mainly is based on the ability of polyelectrolytes to cross-link in the presence of counter-ions to form hydrogels. Ionic biopolysaccharides mainly form meshwork structures by combining with the counter-ions, induce gelation by binding mainly to the polymeric blocks and produce three-dimensional lattices of the ionically cross-linked moiety. Since the use of various ionic biopolysaccharides like alginates, pectin, gellan gum and chitosan for the encapsulation of drugs and even cells, ionic gelation technique has been widely used.

The current book entitled *Ionically Gelled Biopolysaccharide Based Systems in Drug Delivery* covers a thorough insight into the chemistry and mechanism of ionic gelation of various ionic biopolysaccharides like alginate, gellan gum, pectin, chitosan, carboxymethyl cellulose, etc., and the applications of various ionically gelled biopolysaccharides in drug delivery fields. The current book is a collection of nine chapters presenting different key topics related to ionic gelation of various ionic biopolysaccharides emphasizing their drug delivery applications by the leading experts. A concise account on the contents of each chapter has been described to provide a glimpse of the book to the readers.

Chapter 1 entitled "Ionically Gelled Pectinates in Drug Delivery" deals with a comprehensive and functional discussion on the mechanism of ionically gelled pectinates and their potential uses for controlled drug release.

Chapter 2 entitled "Ionically Gelled in Alginates Drug Delivery" presents various applications of ionically gelled alginate-based systems for drug delivery. In addition, biological sources, extraction process and chemistry of alginates along with the mechanism of ionic gelation of alginates have been discussed.

Chapter 3 entitled "Ionically Gelled Gellan Gum in Drug Delivery" highlights important considerations of the ionically gelled gellan gum-based systems and their uses in drug delivery applications.

Chapter 4 entitled "Ionic Gelled Chitosan in Drug Delivery" describes preparations, physicochemical properties and drug delivery applications of various ionically gelled chitosan-based systems.

Chapter 5 entitled "Ionically Gelled Carboxymethyl Polysaccharides in Drug Delivery" outlined various ionically cross-linked carboxymethyl polysaccharide-based carriers for delivery of various drugs, such as chemotherapeutics, anti-inflammatory drugs and protein-based therapeutic agents.

Chapter 6 entitled "Polyelectrolyte Complex-Based Ionically Gelled Biopolymeric Systems for Sustained Drug Release" provides an overview of various aspects of polyelectrolyte complex-based systems prepared by ionic cross-linking gelation of ionic biopolysaccharides and their applications for sustained drug releasing.

Chapter 7 entitled "Ionically Gelled Polysaccharide-Based Interpenetrating Polymer Network Systems in Drug Delivery" summarizes various aspects of polysaccharide-based interpenetrating polymer network systems (IPNs) prepared by ionic cross-linking gelation emphasizing their applications in fabrication of drug delivery platforms.

Chapter 8 entitled "Ionically Gelled Polysaccharide-Based Multiple Units in Drug Delivery" deals the preparation of various drug-releasing multiple-unit systems by ionic cross-linking gelation of ionic biopolysaccharides.

Chapter 9 entitled "Ionically Gelled Polysaccharide-Based Floating Drug Delivery Systems" describes the adaptation of drug delivery system within the gastrointestinal tract by using ionically gelled polysaccharide-based floating drug delivery systems.

We would like to convey our sincere thanks to all the authors of the chapters for providing timely and precious contributions. We thank the publisher—Springer Nature. We specially thank Dr. Vijay Kumar Thakur (Series Editor, Gel Horizons: From Science to Smart Materials, Springer Nature), Swati Meherishi, Silky Abhay Sinha, Vindhya H. Pillai and Ashok Kumar for their helpful support in organization of book editing process from beginning to finishing. We gratefully acknowledge the permissions to reproduce copyright materials from various sources. Finally, we would like to thank our family members, all respected teachers, friends, colleagues and dear students for their continuous encouragements, inspirations and moral supports during preparation of the current book. Together with our contributing authors and the publishers, we will be really delighted if our effort fulfils the needs of academicians, researchers, students, drug delivery formulators, biomedical and pharmaceutical experts.

Mayurbhanj, India — Dr. Amit Kumar Nayak
Daltonganj, India — Dr. Md Saquib Hasnain
Bilaspur, India — Dr. Dilipkumar Pal

Contents

About the Editors

Dr. Amit Kumar Nayak is currently working as an Associate Professor at Seemanta Institute of Pharmaceutical Sciences, Odisha, India. He earned his Ph.D. in Pharmaceutical Sciences from IFTM University, India. He worked as Senior Research Associate at IIT, Kanpur in a CSIR sponsored project. Dr. Nayak has over 12 years of research experiences in the field of pharmaceutics, especially in the development and characterization of novel biopolymeric and nanostructured drug delivery systems made of natural biopolymers. He has authored over 130 research and review articles in various high impact peer-reviewed journals and 102 book chapters in various international books. In addition, he has authored 2 books and edited 10 books published by International publisher(s). He has received University Foundation Day Research Award-2019 by Biju Patnaik University of Technology, Odisha. He serves as a reviewer for several reputed journals and is a life member of Association of Pharmaceutical Teachers of India (APTI).

Dr. Md Saquib Hasnain is currently working as a Professor in the Department of Pharmacy, Palamau Institute of Pharmacy, Jharkhand, India. He has over 9 years of research experience in the field of drug delivery and pharmaceutical formulation analyses, especially systematic development and characterization of diverse nanostructured drug delivery systems, controlled release drug delivery systems, bioenhanced drug delivery systems, nanomaterials and nanocomposites employing Quality by Design approaches and many more. Till date he has authored over 50 publications in various high impact peer-reviewed journals and 100 book chapters to his credit. In addition, he has authored 2 books and edited 9 books published by International publisher(s). He is also serving as the reviewer of several prestigious journals. Overall, he has earned highly impressive publishing and cited record in Google Scholar (H-Index: 26). He has also participated and presented his research work at over ten conferences in India, and abroad. He was also the member of scientific societies i.e., Royal Society of Chemistry, Great Britain, International Association of Environmental and Analytical Chemistry, Switzerland and Swiss Chemical Society, Switzerland.

Dr. Dilipkumar Pal (born in W.B., India) Ph.D., M. Pharm, Chartered Chemist, Post Doct (Australia) is an Associate Professor in the Department of Pharmaceutical Sciences, Guru Ghasidas Vishwavidyalaya (A Central University), Bilaspur, C.G., India. He received his master and PhD degree from Jadavpur University, Kolkata and performed postdoctoral research as "Endeavor Post-Doctoral Research Fellow" in University of Sydney, Australia. His areas of research interest include "Isolation, structure Elucidation and biological evaluation of indigenous plants", "Synthetic Chemistry and pharmacological evaluations" and "natural biopolymers". He has published 172 full research papers in peer-reviewed reputed national and international scientific journals, having good impact factor and contributed 114 abstracts in different national and international conferences. He has written onebook, forty-nine book chapters and edited five books published by reputed international publishers. His research publications have acquired a highly remarkable cited record in Scopus and Google Scholar (H-Index: 40; i-10-index 96, total citations 5239 as on date). Dr. Pal, in his 21 years research-oriented teaching profession, received 13 prestigious national and international professional awards also. He has guided 8 PhD and 40 master students for their dissertation thesis. Dr Pal's name has been included in world ranking of top 2 % Indian Scientist. He is the reviewer and Editorial Board member of 27 and 29 scientific journals, respectively. Dr. Pal has been working as the Editor-in-Chief of one good research journal also. He is the member and life member of fifteen professional organizations.

Chapter 1
Ionically Gelled Pectinates in Drug Delivery

Amit Kumar Nayak, Md Saquib Hasnain, and Dilipkumar Pal

Abstract Pectins are hydrophilic linear anionic polysaccharide extracted from plant cell walls. These are inexpensive, biodegradable and biocompatible in nature. Pectins are being employed as additives, thickeners and gelling agents in many foods, cosmetics and pharmaceutical applications. Since past few decades, pectins have widely been investigated for its unique nature of forming hydrogels (mainly, low methoxy pectin) by the influence of various divalent metal ions. These ionically gelled pectinate hydrogels are being investigated as potential carriers in controlled release delivery of many drugs. The current chapter deals with a comprehensive and functional discussion on the mechanism of ionically gelled pectinates and their potential uses for controlled drug release.

Keywords Pectin · Cross-linking · Ionic gelation · Hydrogels · Drug delivery

1 Introduction

At present, the socio-economic condition of the modern world has elevated the interest of natural polymers [1, 2]. Usually, low manufacturing cost of natural polymers related to their large availability in nature is recognized as the main advantage for their extensive uses in almost all aspects of daily uses [3, 4]. Besides these, biocompatibility and biodegradability are the two important advantages that favor their biomedical applications including delivery of drugs, proteins and peptides, tissue engineering, wound dressing, orthopaedics, dentistry, etc. [1, 5–8]. Among various

A. K. Nayak
Department of Pharmaceutics, Seemanta Institute of Pharmaceutical Sciences, Jharpokharia, Mayurbhanj 757086, Odisha, India

M. S. Hasnain (✉)
Department of Pharmacy, Palamau Institute of Pharmacy, Chianki, Daltonganj 822102, Jharkhand, India

D. Pal
Department of Pharmaceutical Sciences, Guru Ghasidas Vishwavidyalaya, Koni, Bilaspur 495009, Chhattisgarh, India

A. K. Nayak et al. (eds.), *Ionically Gelled Biopolysaccharide Based Systems in Drug Delivery*, Gels Horizons: From Science to Smart Materials,
https://doi.org/10.1007/978-981-16-2271-7_1

natural polymers, natural polysaccharides are the most popular biopolymeric group, which are extensively used for the designing of various biopolymer-based matrices for drug delivery [9–11].

Polysaccharides are the polymers of monosaccharides. Natural polysaccharides are extracted from various natural origins: plant origin (e.g., pectin, guar gum), microbial origin (e.g., dextran, xanthan gum), algal origin (e.g., alginate), and animal origin (e.g., chitosan, chondroitin) [12–18]. Polysaccharides, in particular, possess some excellent properties, which make these as the biopolymeric group with the widest biomedical applications [19–21]. Important properties of natural polysaccharides are non-toxicity (monomer residues are not hazardous to health), water solubility, high swelling ability, stability to pH variations, a broad variety of chemical structures, ease of chemical modifications, etc. [22–24]. Particularly, most of the natural polysaccharides have hydrophilic groups such as hydroxyl, carboxyl, and amino groups, which could form non-covalent bonds with biological tissues (mainly epithelial and mucous membranes), forming bioadhesion [25]. Based on the electrical or charge nature, these polysaccharides can be classified into 2 classes [26, 27]: (i) polyelectrolyte polysaccharides and (ii) non-polyelectrolyte polysaccharides. Non-polyelectrolyte polysaccharides include guar gum, locust bean gum, tamarind gum, arabinans, xanthan gum, amylase, cellulose, etc. [28]. However, polyelectrolyte polysaccharides can be sub-classified into two distinct sub-classes [26, 29]: (i) negatively charged (anionic) polysaccharides (e.g., alginate, pectin, gellan gum, heparin, etc.) and positively charged (cationic) polysaccharides (e.g., chitosan). These polyelectrolyte polysaccharides usually undergo ionic cross-linking gelation (or ionic gelation) in the presence of counterions to form polysaccharide-based hydrogels [30–42]. Over the last few years, a great deal of attention has been paid to the development of polysaccharide-based hydrogels using ionic gelation of various polyelectrolyte polysaccharides [43–53]. Likewise various anionic natural polysaccharides, pectin (a hydrophilic linear anionic polysaccharide extracted from plant cell walls) has been investigated widely for its unique nature of forming hydrogels (mainly, low methoxy pectin) by the influence of various divalent metal ions, and these ionically gelled pectinate hydrogels are being investigated as potential carriers in controlled release delivery of many drugs [29, 53–55]. The current chapter deals with a comprehensive and functional discussion on the mechanism of ionically gelled pectinates and their potential uses for controlled drug release.

2 Ionic Gelation of Natural Polysaccharides

Ionic gelation of polysaccharides is occurred based on the ionic interaction-based cross-linking capability of polyelectrolyte polysaccharides by the influence of various divalent or trivalent metal ions [29]. In ionic gelation technique, solutions of ionic polysaccharides (alginate, pectin, gellan gum, etc.) are added drop wise to the aqueous solutions containing counterions [26]. Due to electrostatic interaction between oppositely charged species, ionic polysaccharides undergo ionic cross-linking gelation and precipitate to form gelled polysaccharidic matrices [27]. The ionic gelation technique is very simple and the conditions used were very mild [56]. Additionally, physical cross-linking because of ionic cross-linking gelation instead of chemical cross-linking gelation avoids the possible toxicity due to reagents and other unwanted outcomes [26, 27]. These ionic polysaccharides form the meshwork structures by combining with the counterions and induce gelation by binding mainly to the polymeric blocks and produces three-dimensional lattices of ionically crossed-linked moiety [30]. The ionic gelation is dependent on several factors like polymer-type, polymer concentration, cross-linker type, cross-linker concentration, pH of the cross-linking solutions, etc. [29, 57]. Since the use of various ionic polysaccharides like sodium alginate [30], low methoxy pectin [55, 58], gellan gum [48–51], carboxymethyl cellulose [59, 60], chitosan [61], etc., for the encapsulation of drugs, ionic gelation technique has been widely used.

The counterions used for ionic gelation of various ionic polysaccharides can be categorized into two categories [26]: low molecular weight counterions (e.g., $CaCl_2$, $BaCl_2$, $MgCl_2$, $CuCl_2$, $ZnCl_2$, $CoCl_2$, Al_2Cl_3, pyrophosphate, tripolyphosphate, tetrapolyphosphate, octapolyphosphate, hexametaphosphate; and high molecular weight ions (e.g., octylsulphate, laurylsulphate, hexadecylsulphate, cetylstearylsulphate). In another way, they can be classified as [26, 29]: anionic cross-linkers for cross-linking of cationic polymers (e.g. pyrophosphate, tripolyphosphate, tetrapolyphosphate) and cationic cross-linkers for cross-linking of anionic polymers (e.g., $CaCl_2$, $BaCl_2$, $MgCl_2$, $CuCl_2$, $ZnCl_2$, $CoCl_2$, Al_2Cl_3).

Since the uses of various ionic polysaccharides like alginates, pectin, gellan gum, carboxymethyl cellulose, and chitosan for the encapsulation of drugs, ionic gelation technique has been widely used [27, 30]. The natural ionic polysaccharides in spite, having a property of coating on the drug core and act as release rate retardants contains certain anions on their chemical structure. In different drug delivery researches, various categories of drugs have been successfully incorporated within ionically gelled natural polysaccharide-based hydrogel matrices with different patterns of drug entrapments and release profiles depending on their method of preparations, compositions, nature of ionic cross-linking, physicochemical properties of excipients and incorporated drugs.

3 Pectin and Ionic Gelation Mechanism of Pectins

Pectins are inexpensive, non-toxic, water soluble, natural polysaccharide extracted industrially from citrus peels, sugar beet roots, apple pomaces, etc. and have widely been employed as additives, thickeners and gelling agents in many foods, cosmetics and pharmaceutical applications [62–64]. Pectin consists of linearly connected (1–4) linked α-D-galacturonic acid residues interrupted by some rhamnogalacturonic acid residues and α-L-rhamnopyranose by α-(1–2) linkage (Fig. 1) [65].

The galacturonic acid residue of the pectin backbone is partially esterified. Pectins are generally characterized by degree of esterification and degree of amidation (in

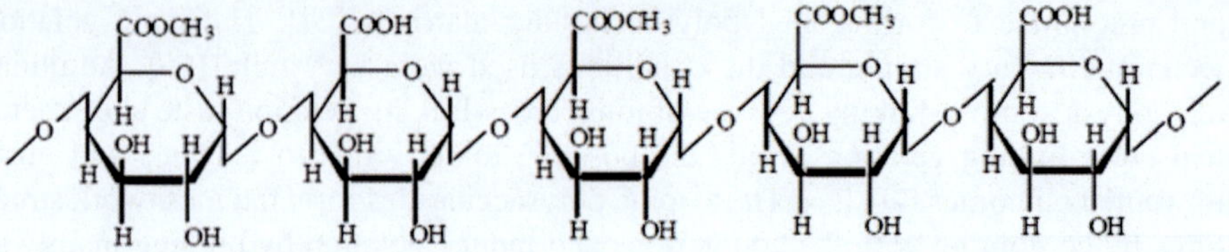

Fig. 1 Structure of pectin (own drawing)

some cases), which both are regarded as a fraction of esterified and amidated carboxyl groups, respectively [66]. Pectins can be classified as low methoxypectins (LMPs, with a 25–50% degree of methylation) and high methoxypectins (HMPs, with a 50–80% degree of methoxylation) pectins [67].

The degree of esterification has a major influence on pectin properties, including solubility and gel-forming ability [68, 69]. HMPs produce gel in an acidic medium on addition of a high amount of sucrose (>50%) [62]. Usually, LMPs can form gel structures by ionic gelation with various divalent metal cations (e.g., Ca^{2+}, Zn^{2+}, etc.) for the use as effective vehicles in drug delivery applications [55, 70, 71]. The interactions of the carboxyl groups of LMP backbone with divalent ionic cross-linking cations induce the formation of the so-called Egg-Box structure, even though it slightly differs from the "Egg-Box" model originally defined for ionotropic gelation of sodium alginate [62]. According to the "Egg-Box" model originally hypothesized for LMP, there is an initial dimerization step of two homogeneous galacturonic chains by cooperative bridging of parallel facing chains through divalent ionic cross-linking cations [26]. Considerately, it is possible as a result of the fact that the homogalacturonic chains are quite rigid and binding of first ionic cross-linking metal cations by two pectin chains assists their alignment with regard to each other, which permits the easier binding in a next chain, and so on along the sequence [72, 73]. The anti-parallel orientation of the two pectin chains appears to be the most favorable arrangement, and besides the electrostatic interactions, this initial dimer association is strongly stabilized by the hydrogen bonding [26]. Subsequent ionic gelation-induced aggregation of the "Egg-Box" dimmers in tetramers, hexamers, etc. can occur; but these subsequent associations of dimmers have no particular specificity and seem to be merely governed by electrostatic interactions [62]. The multimers are therefore easily disrupted by competing monovalent ions, which is initially formed chain dimmers are not.

4 Drug Delivery Applications of Ionically Gelled Pectins

Ionically cross-linked pectinate hydrogels, formed by Ca^{2+} or Zn^{2+}-induced ionic gelation of low methoxyl pectin, have been regarded as an excellent drug delivery matrices because of their unique features, such as nontoxic, highly biocompatible, mechanically strong and acid-stable characteristics [54, 55, 67]. These pectinate hydrogels are generally prepared by the so-called ionic gelation technique in which low methoxy pectin solutions containing drugs are extruded drop wise into divalent metal ion containing solutions to form beads [74]. During last few decades, numerous ionically gelled pectinate microparticles/beads for oral drug delivery have been investigated [27, 53–55]. Various research groups have investigated the effect of pectin types (degree of methylation), preparation methodologies, drying conditions, cross-linking time, pH of cross-linking solutions, cross-linker concentration, etc., on drug encapsulation, drug release, swelling, morphology, etc. [53–55, 67, 69–71].

In a research, Das and Ng [54] prepared ionically gelled calcium pectinate beads of delayed release of resveratrol by varying various formulation parameters like concentration of resveratrol, concentration of pectin, concentration of ionic cross-linker, cross-linking period, cross-linking pH and drying condition of prepared beads [54]. Most of the resveratrol-loaded calcium pectinate beads were found spherical in shape with 1 mm of diameter (approximately). The in vitro swelling and the drug retention pattern were found to be influenced with different formulation variables. The in vitro drug release results exhibited the linearity of plots for the cumulative percent of resveratrol release versus the square root of release period often after an initial lag period (for almost all these resveratrol-loaded calcium pectinate beads). The optimized formulation of resveratrol-loaded calcium pectinate beads was found to encapsulate high resveratrol contents (>97.5%). The in vitro resveratrol releasing from these resveratrol-loaded calcium pectinate beads in the intestinal media after 10 h was found 80–100%. The overall results of this investigation demonstrated that these resveratrol-loaded calcium pectinate beads can be employed for the site-specific delayed drug releasing in the lower gastrointestinal tract.

Jantrawut et al. [75] developed rutin-loaded ionically gelled beads made of low methoxyl pectin [75]. The mechanical characterization results demonstrated that that the Young's modulus of non-amidated LMP gels was found to be decreased with the addition of 15% sorbitol in the bead formula. The particle sizing of rutin-loaded ionically gelled calcium pectinate beads was 600 μ, approximately, and the morphology of these beads was found to be oblonged shape with dense matrix. These beads were characterized by scanning electron microscopy (SEM) and these micrographs are presented in Fig. 2. The surface of the sphere of rutin-loaded calcium pectinate beads made of 3% non-amidated low methoxy pectin exhibited a rough and globulous morphological feature (Fig. 2a), while rutin-loaded calcium pectinate beads made of 3% non-amidated low methoxy pectin and 15% sorbitol presented a much smoother surface (Fig. 2c). The rutin-loaded calcium pectinate beads contained sodium carbonate ($NaHCO_3$) exhibited oblonged shape with wrinkled morphology (Fig. 2b, d).

The cross-sectional micrographs of all these rutin-loaded calcium pectinate beads presented denser and small hollow pockets within their matrices, especially for rutin-loaded calcium pectinate beads contained $NaHCO_3$. This occurrence might be attributed by the fact of gas bubbles formation by the gas-forming agent (here $NaHCO_3$), which might increase the bead porosity. The rutin-loaded ionically gelled calcium pectinate beads containing sodium bicarbonate exhibited the rutin encapsulation efficiency less than 80% (approximately), which was found to be lowered with

Fig. 2 SEM images of calcium pectinate beads made of 3% non-amidated low methoxy pectin **a** calcium pectinate beads made of 3% non-amidated low methoxy pectin, 15% sorbitol and 1% bicarbonate **b** calcium pectinate beads made of 3% non-amidated low methoxy pectin and 15% sorbitol **c** and calcium pectinate beads made of 3% non-amidated low methoxy pectin and 1% bicarbonate **d** (*left*) and their cross-sections of respective bead (*right*) at magnification × 150 [75] (with permission, Copyright © 2013 Elsevier Ltd.)

the increasing of pHs of the cross-linking solutions used for the bead preparation. The rutin-loaded calcium pectinate beads made of 3% non-amidated low methoxy pectin and 15% sorbitol exhibited the favorable in vitro release as well as swelling characteristics. The results of the study indicated that the gel-texture has influenced the release of rutin from these ionically gelled calcium pectinate beads.

Ghibaudo et al. [76] developed iron pectinate beads by ionic cross-linking gelation to deliver iron to the intestinal cells [76]. In this study, spherically shaped ionically gelled iron pectinate beads having 1–2 mm diameter were produced. SEM image of these beads indicated a rather smooth external surface morphology with an inner structural feature showing a compact texture and occurrence of some holes on the bead surface (Fig. 3).

These ionically gelled iron pectinate beads exhibited high density of 1.29 g/mL and high porosity of 93.28% at low pressure, suggesting their higher permeability even when low pressure was applied. The in vitro swelling in simulated intestinal medium (pH 8) was found elevated in comparison to that of in simulated gastric medium (pH 1.2). The physiological assays were performed by exposing these ionically gelled iron pectinate beads within simulated gastric as well as intestinal milieu to measure the physiological absorption of iron and the trans-epithelial transport of iron from these beads into the Caco-2/TC7 cells. The sources of iron as ferrous sulfate (control) or ionically gelled iron pectinate beads did not have shown any significant action on the absorption of iron by the Caco-2/TC7 cells. Concerning the transport of iron, the iron contents attached to the apical pole of Caco-2/TC7 cell line monolayers was recovered in the basal compartment, which was found to be proportional with the exposure period. After incubation for a period of 4 h, the iron transport from these ionically gelled iron pectinate beads was found significantly elevated in comparison to that of the iron from ferrous sulfate (control) (Fig. 4).

The overall results of this study indicated that these ionically gelled iron pectinate beads can be used to surmount the lower competence of iron transport devoid of varying the sensory characteristics.

Chaurasia et al. [77] developed methotrexate-loaded ionically gelled calcium pectinate microspheres for colon-specific delivery [77]. These methotrexate-loaded ionically gelled calcium pectinate microspheres were prepared *via* modified emulsification technique employing calcium cations as ionic cross-linking ions. Particle sizing of the microspheres was measured by laser diffraction size analyzer, which was found within the range, 20.82 ± 1.34 to 32.26 ± 1.59 μ. Methotrexate encapsulation efficiency of these prepared calcium pectinate microspheres was measured by enzyme pectinase-induced digestion for 24 h, which was found within the range, 52.28 ± 0.32 to 74.01 ± 3.32%. In vitro releasing of methotrexate from these methotrexate-loaded ionically gelled calcium pectinate microspheres was measured in simulated gastric fluid for 2 h and then, in simulated intestinal fluid for 3 h. In vitro releasing of methotrexate in simulated colonic fluid in presence of rat caecal content was also carried out. In addition, in vitro methotrexate releasing rate was also measured after the enzymatic induction by the treatment of the rats using 1 ml aqueous dispersion of pectin (1% w/v) for a period of 7 days. In vitro releasing of methotrexate from

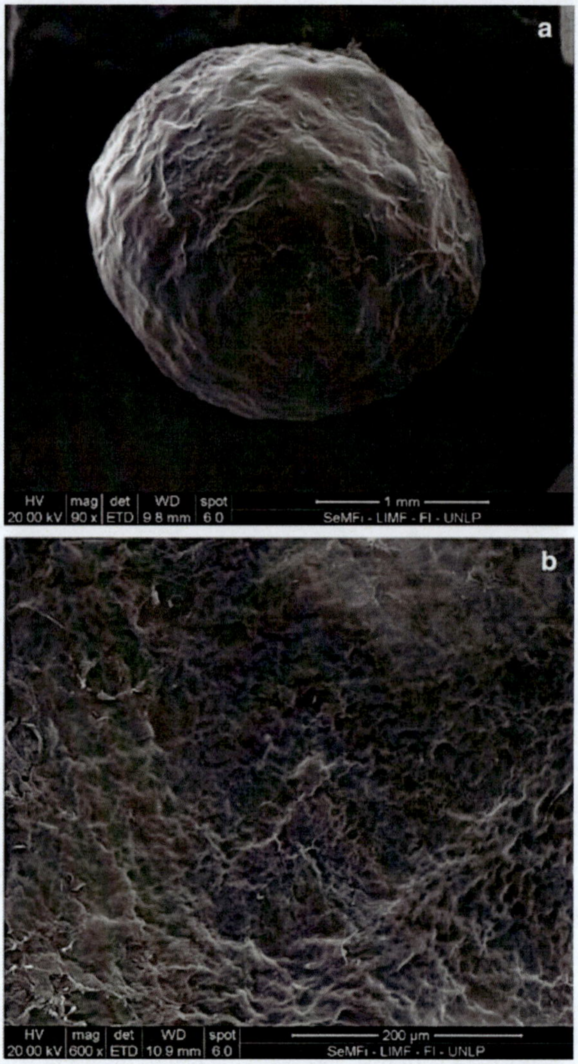

Fig. 3 SEM images of ionically gelled iron pectinate beads: **a** External surface and **b** Inner structure [76] (with permission, Copyright © 2018 Elsevier B.V.)

these methotrexate-loaded ionically gelled calcium pectinate microspheres in simulated gastric fluid as well as simulated intestinal fluid presented that only 8.15 ± 0.49% of loaded methotrexate was released within 5 h, whereas the majority of the loaded methotrexate was measured to be released in pectinase-contained simulated colonic fluid. In vitro methotrexate releasing results exhibited that the releasing of 69.94 ± 3.46% of loaded methotrexate in presence of 3% of rat caecal content, which was noticed to be further augmented to 94.43 ± 4.48% when enzymatic induction was performed for a period of seven days. Therefore, the overall results of this

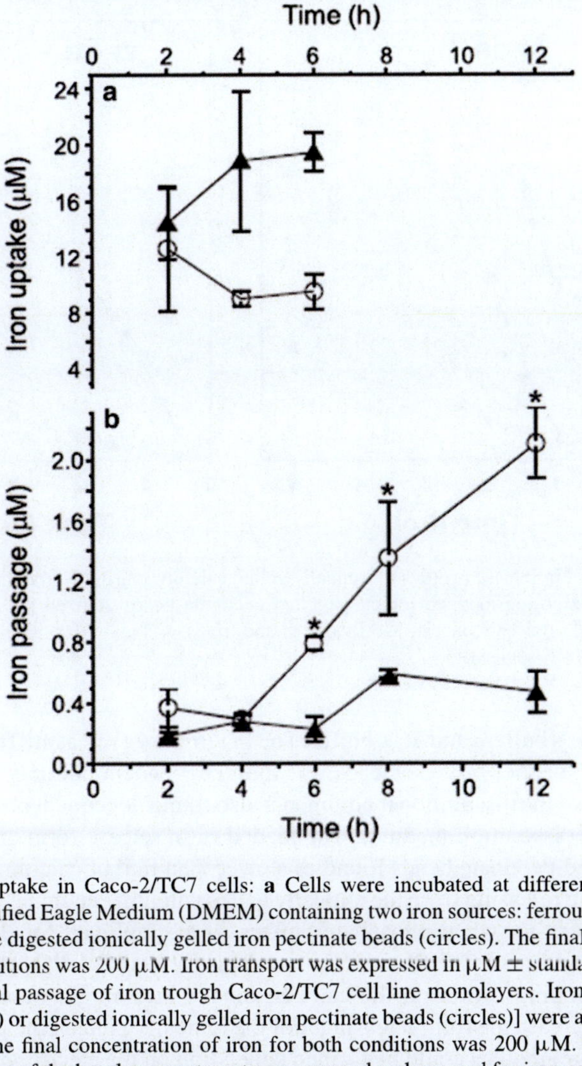

Fig. 4 **a** Iron uptake in Caco-2/TC7 cells: **a** Cells were incubated at different time points in Dulbecco's Modified Eagle Medium (DMEM) containing two iron sources: ferrous sulfate (control, triangles) and the digested ionically gelled iron pectinate beads (circles). The final concentration of iron in both conditions was 200 μM. Iron transport was expressed in μM ± standard deviation, and **b** Trans-epithelial passage of iron trough Caco-2/TC7 cell line monolayers. Iron sources [ferrous sulfate (triangles) or digested ionically gelled iron pectinate beads (circles)] were added to the upper compartment. The final concentration of iron for both conditions was 200 μM. At different time points, the content of the basal compartment was removed and assayed for iron concentration. Iron transport was expressed in μM ± standard deviation. Asterisks indicate significant differences ($p < 0.05$) between the iron concentrations in the basal compartment for the two treatments at the same time point [76] (with permission, Copyright © 2018 Elsevier B.V.)

study clearly concluded that methotrexate-loaded ionically gelled calcium pectinate microspheres can be used for site-specific delivery of drugs in colonic region.

In a work by Assifoui et al. [66], rutin-loaded ionically gelled calcium pectinate beads and zinc pectinate beads were formulated for colonic delivery [66]. The influences of the dissolution medium on the in vitro drug release from these ionically gelled pectinate beads were evaluated in 3 different dissolution medium: Sorensen's

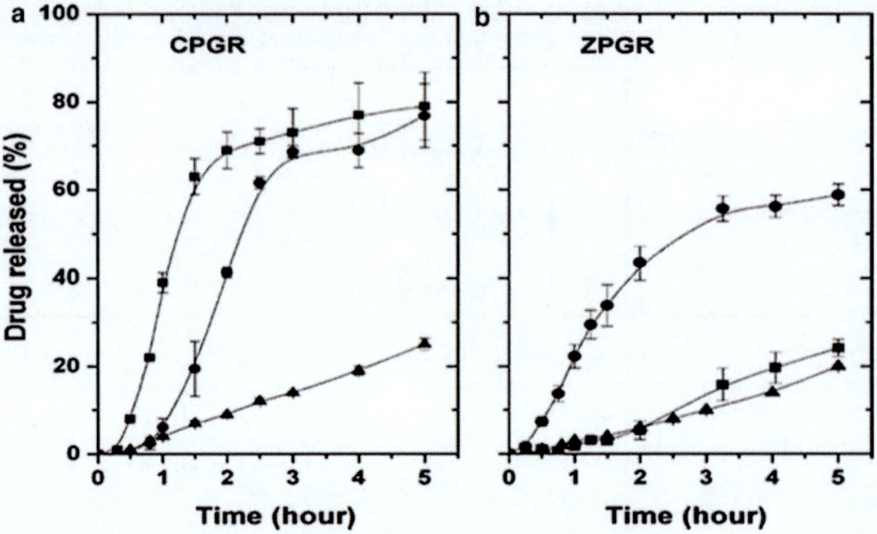

Fig. 5 In vitro rutin release profiles of ionically gelled calcium ion-induced ionically gelled pectinate beads and zinc ion-induced ionically gelled pectinate beads at fixed pH 7.3 in 3 different buffers: square, Sorensen's; circle, Mc Ilvaine's; and triangle, Tris-buffer. [66] (with permission, Copyright © 2011 Elsevier Ltd.)

and Mc Ilvaine's buffers and Tris-buffer. The in vitro drug releasing from these beads and swelling behavior of these beads was found dependent not only on the pHs and ionic strength of the dissolution medium, but also found dependent on the electrolytes present in the dissolution medium. The loaded rutin release form zinc ion-induced ionically gelled pectinate beads found as slower than that of calcium ions (Fig. 5).

In vitro rutin releasing from the ionically gelled calcium pectinate beads was found comparatively faster when phosphate buffers were employed for drug dissolution due to the formation of $CaHPO_4$ precipitate. This formed $CaHPO_4$ precipitate could produce a pumping action on the calcium ions, which might destabilize the gel-structure to increase the rutin releasing. For the ionically gelled zinc pectinate beads, two types of precipitates could be formed depending on the electrolytes composition of the dissolution medium used for in vitro dissolution testing. The formation of $Zn_3(PO_4)_2$ and its coat onto the beads might reduce the in vitro rutin releasing (in Sorensen's buffer). In contrast, the formation of $ZnHPO_4$ might produce the pumping action of zinc ions released from the ionically gelled zinc pectinate beads, which augmented the in vitro rutin releasing (in Mc Ilvaine's buffer).

Dhalleine et al. [78] also investigated in vitro and in vivo theophylline release characteristics of ionically gelled calcium pectinate beads and zinc pectinate beads loaded with theophylline [78]. The measured bead sizing was found to be varied in between 1.8–2.8 mm and the theophylline encapsulation was measured in between 27–30% for theophylline-loaded calcium pectinate beads and zinc pectinate beads, respectively. All these ionically gelled pectinate beads were evaluated for in vitro theophylline releasing in different release media (water, phosphate buffer, and Tris-buffer at pH 7.4). The in vitro theophylline releasing profiles of zinc pectinate beads loaded with theophylline was found to be very much dependent on the release medium used for testing, while calcium pectinate beads loaded with theophylline exhibited a comparative faster disintegration. The in vitro theophylline releasing from zinc pectinate beads loaded with 20% and 40% theophylline was found to be sustained when evaluated in phosphate buffer (Fig. 6) and simulated colonic medium (Fig. 7), in the occurrence of phosphate ions.

In vivo pharmacokinetic profiles of these theophylline-loaded pectinate beads were tested in rats. The *in vivo* results of zinc pectinate beads loaded with theophylline led to a significant lag time (t_{max} of 12.0 ± 0.1 h) for the theophylline absorption in comparison to that of calcium pectinate beads loaded with theophylline (t_{max} of 6.0 ± 2.8 h) and free theophylline (t_{max} of 2.5 ± 2.1 h).

Das et al. [55] developed zinc ion-induced ionically gelled pectinate beads for delayed release of resveratrol for the treatment of lower gastrointestinal diseases [55]. These zinc pectinate beads exhibited higher resveratrol encapsulation. Almost all these formulated resveratrol-loaded zinc pectinate beads were of spherically shaped having 1 mm of bead diameter (approximately). The in vitro drug release results demonstrated a better delayed resveratrol releasing profile as compared to that of calcium pectinate beads. The resveratrol retention and swelling erosion characteristics were greatly influenced by various tested formulation parameters. These resveratrol-loaded zinc pectinate beads were found to be well fitted with zero-order

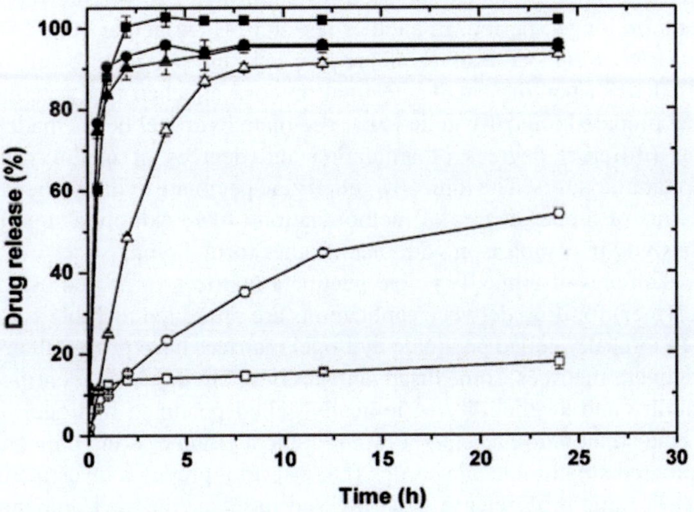

Fig. 6 In vitro dissolution profiles of theophylline from calcium pectinate beads loaded with theophylline (full symbols) and zinc pectinate beads loaded with theophylline (empty symbols) in three different media: phosphate buffer (square), water (circle), and Tris-buffer (triangle) at pH 7.4 [78] (with permission, Copyright © 2011 Elsevier B.V.)

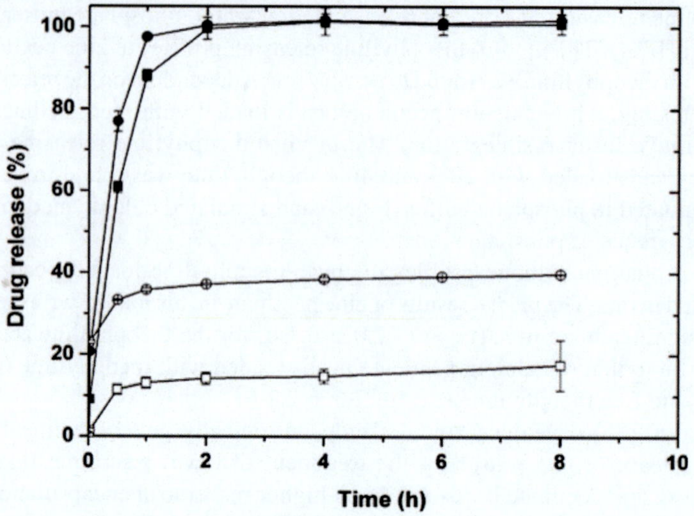

Fig. 7 In vitro dissolution profiles of theophylline from calcium pectinate beads loaded with theophylline (full symbols) and zinc pectinate beads loaded with theophylline (empty symbols) in phosphate buffer (square) and phosphate buffer containing pectinolytic enzymes (circle) [78] (with permission, Copyright © 2011 Elsevier B.V.)

kinetic model. The results of this study demonstrated that the optimized resveratrol-loaded zinc pectinate beads can be used for controlled delayed delivery of resveratrol over a prolonged period. In another research, Hagesaether et al. [53] reported excellent *ex vivo* mucoadhesivity of pre-swelled zinc pectinate beads tested onto the inverted fresh porcine small intestinal mucosa attached to a rotating cylinder [53]. They prepared ionically gelled zinc pectinate hydrogel beads made of 6 kinds of pectins (different degrees of amidation and degrees of methoxylation) from different manufacturers. The ionically gelled zinc pectinate hydrogel beads prepared using pectins of higher degree of methoxylation (70%) exhibited superior *ex vivo* mucoadhesivity in comparison with that of other formulations.

Some examples of ionically gelled pectinate matrices (prepared using pectin as single polymer) for drug delivery applications are presented in Table 1.

Though ionically gelled pectinate hydrogel matrices have proved their capability as drug delivery matrices, some disadvantages have already been identified [69, 74]. The solubility and swellability of ionically gelled pectinate hydrogel matrices in the gastrointestinal fluid presented low entrapment efficiency and premature release of incorporated small molecular drugs [58, 80]. In a previous investigation by Das and Ng [54], rapid drug release was observed in the acidic environment following the exposure to the alkaline [54]. In another research, Atyabi et al. [86] reported similar observation of enhanced drug release after consecutive exposure to acidic and basic environment [86]. To overcome these drawbacks, several modification of ionically gelled pectinate hydrogel matrices have been investigated by various research groups [65, 71, 74, 94–97]. Among various modifications of ionically gelled

Table 1 Ionically gelled pectinate matrices (prepared using pectin as single polymer) for drug delivery applications

Ionically gelled pectinate matrices	Drugs incorporated	References
Calcium pectinate beads	Indomethacin	Chung and Zhang [79]
Calcium pectinate gel beads	Catechin	Lee et al. [80]
Calcium pectinate beads	Metronidazole	Pawar et al. [81]
Zinc pectinate gel microparticles	Ketoprofen	El-Gibaly [82]
Calcium pectinate gel beads	Bovine serum albumin	Sriamornsak [83]
Hollow/porous calcium pectinate floating beads	Diclofenac sodium	Badve et al. [84]
Calcium pectinate beads	Metronidazole	Sriamornsak et al. [85]
Calcium pectinate beads	Bovine serum albumin	Atyabi et al. [86]
Calcium pectinate gel beads	Theophylline anhydrous	Sriamornsak et al. [70]
Calcium pectinate beads	5-fluorouracil	Jain et al. [87]
Calcium pectinate gel microbeads	Insulin	Si et al. [88]
Calcium pectinate nanoparticles	Insulin	Chang and Lin [89]
Calcium pectinate beads and zinc pectinate beads	Ketoprofen	Dupuis et al. [90]
Zinc pectinate beads	Pterostilbene	Ansari et al. [91]
Zinc pectinate beads	Ketoprofen	Chambin et al. [92]
Zinc pectinate microparticles	Mesalazine	Kawadkar et al. [93]

pectinate hydrogels, use of polymer-blends with low methoxy pectin is regarded as one of the popular and simple approach to prevent premature release of incorporated drugs [58, 69, 74, 98, 99].

In a research by Das et al. [100], zinc pectinate-chitosan composite microparticles of resveratrol were developed and investigated for colon-specific drug delivery [100]. The resveratrol loading and resveratrol encapasulation efficiency were measured as 17.82–48.31% and 96.95–98.85%, respectively. These resveratrol-loaded microparticles were of spherically shaped with 920.48–1107.56 μm of particle sizing. The weight loss and moisture contents of these microparticles were measured as 89.83–94.34% and 8.31–13.25%, respectively. Various formulation parameters (molecular weight of chitosan, concentration of chitosan, concentration of resveratrol, cross-linking period and pH of cross-linking medium) exhibited significant effects on the in vitro drug (here resveratrol) release profiles. Zinc pectinate-chitosan composite microparticles of resveratrol prepared at pH 1.5 using 1% chitosan, pectin to drug ratio of 3:1, and cross-linking period of 2 h, presented colon-specific release of resveratrol. *In vivo* pharmacokinetic testing was carried out for zinc pectinate-chitosan composite microparticles of resveratrol in rats and the results were compared with that of the zinc pectinate microparticles of resveratrol. The overall *in vivo* results indicated the potential of colon-specific drug releasing from these ionically gelled inc pectinate-chitosan composite microparticles of resveratrol, *in vivo* (Fig. 8).

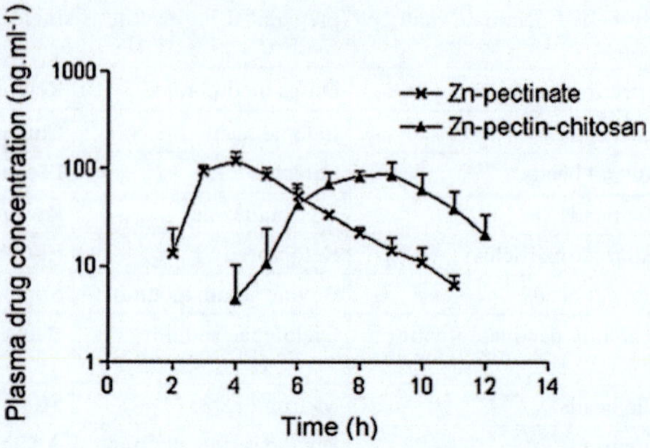

Fig. 8 Plasma concentrations of drug after a single dose of zinc pectinate microparticles of resveratrol (Group 1) and zinc pectinate-chitosan composite microparticles of resveratrol (Group 2). The lines represent the predicted values. Symbols represent the mean observed values ± standard deviation ($n = 6$) [100] (with permission, Copyright © 2010 Elsevier B.V.)

Maestrelli et al. [101] formulated ionically gelled calcium pectinate-chitosan beads and zinc pectinate-chitosan beads for colon-specific delivery of theophylline [101]. In this work, divalent calcium and zinc cations were used for ionic cross-linking. Zinc pectinate-chitosan beads of theophylline exhibit a stronger gelled network as compared to that of the calcium pectinate-chitosan beads of theophylline, enabling higher drug entrapment efficiency, and this was further augmented with the increment of chitosan content in the bead formula, which could possibly be because of the formation of polyelectrolyte complexes in between anionic natured pectin and cationic natured chitosan. The drug transport across the Caco-2 cells presented a significant ($p > 0.05$) increment of drug (here theophylline) permeation from all these ionically gelled pectinate beads concerning theophylline alone. This occurrence could be attributed by the fact of permeation enhancer properties and/or mucoadhesion characteristics of the constituent polymers (pectin and chitosan). The overall drug permeation results demonstrated that the calcium pectinate-chitosan beads of theophylline were found more effective than the zinc pectinate-chitosan beads of theophylline. These ionically gelled pectinate beads of theophyllene were processed to be formulated as enteric-coated tablets and these tablets exhibited excellent colon-specific properties. However, there was no difference in theophyllene releasing from various ionically gelled calcium pectinate-chitosan beads and zinc pectinate-chitosan beads of theophyllene, in vitro.

Rezvanian et al. [102] prepared simvastatin-loaded ionically gelled hydrogel films made of alginate-pectin using ionic cross-linker solutions containing varied calcium chloride concentrations (0.5–3% w/v) and cross-linking periods (2–20 min) [102]. The preparation mechanism of simvastatin-loaded alginate-pectin hydrogel films and the process of ionic cross-linking for hydrogel formation is presented in Fig. 9.

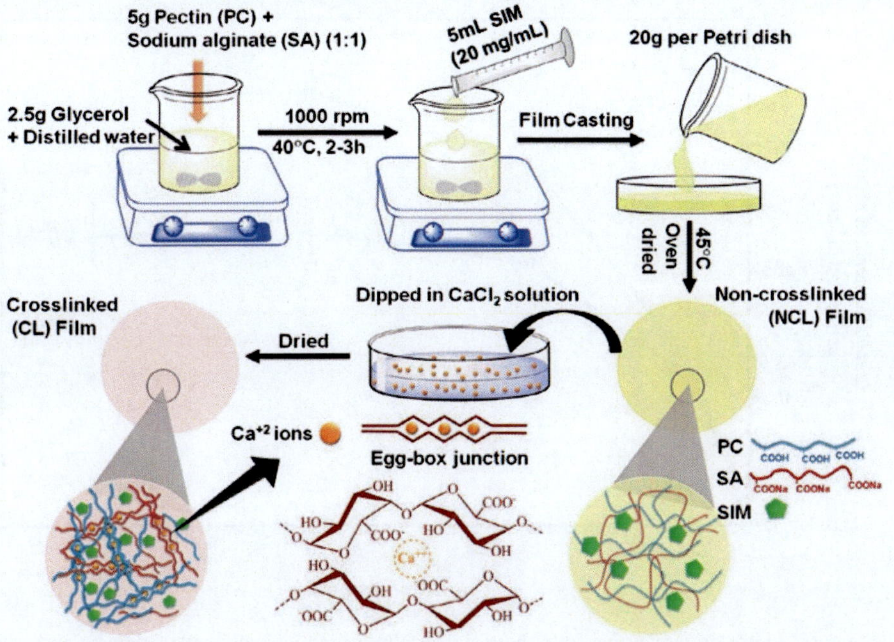

Fig. 9 Preparation mechanism of simvastatin-loaded ionically gelled alginate-pectinate hydrogel films and cross-linking process [102] (with permission, Copyright © 2017 Elsevier B.V.)

In the formulation of simvastatin-loaded hydrogel films, in this work, the ionic cross-linking degree of alginate and pectin were influenced by ionic cross-linker concentrations and cross-linking periods. The preformulation study was carried out to find optimized formula and the optimized simvastatin-loaded ionically gelled alginate-pectinate hydrogel films were prepared using 0.5% and 1% w/v calcium chloride as ionic cross-linker for cross-linking period 2 min. The simvastatin-loaded ionically gelled alginate-pectinate hydrogel films formulated using 1% w/v calcium chloride for cross-linking period 2 min were of rigid and inflexible in nature. This occurrence could be owing to the higher degree of ionic cross-linking. On the other hand, the use of 0.5% w/v calcium chloride for cross-linking period more than 2 min produced the simvastatin-loaded ionically gelled alginate-pectinate hydrogel films of flexible nature with smooth edges and this formula for ionically gelled alginate-pectinate hydrogel films was selected for further study. In vitro simvastatin releasing from these ionically gelled alginate-pectinate hydrogel films was performed using Franz diffusion cell and the results indicated a steady slower releasing of loaded simvastatin from these films. Though the alginate-pectinate hydrogel films formulated without cross-linking demonstrated a relatively faster release of loaded simvastatin, the ionically cross-linked alginate-pectinate hydrogel films exhibited a more sustained releasing of loaded simvastatin, in vitro (Fig. 10).

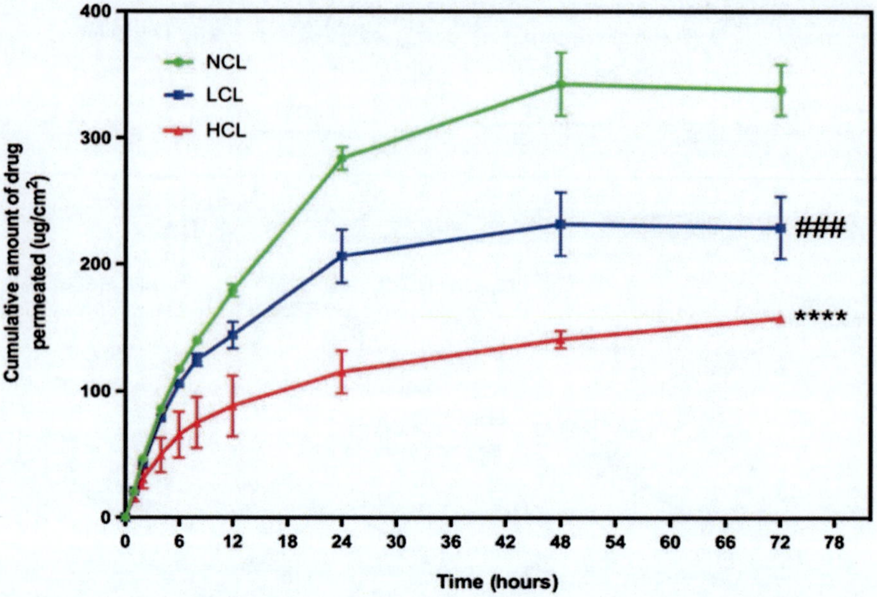

Fig. 10 In vitro drug releasing from these simvastatin-loaded ionically gelled alginate-pectinate hydrogel films using Franz diffusion cell (NCL: non-cross-linked, LCL: low cross-linked, HCL: high cross-linked) [102] (with permission, Copyright © 2017 Elsevier B.V.)

For biocompatibility evaluation, these ionically gelled hydrogel films were tested using human skin fibroblasts and the results of in vitro cytotoxicity clearly indicated the non-toxicity of these hydrogel films for their possible use in wound dressing applications.

Ionically gelled pectinate-tamarind seed polysaccharide mucoadhesive beads were developed for intestinal mucoadhesive delivery of metformin HCl *via* oral administration [69]. For the formulation of these ionically gelled pectinate-based mucoadhesive beads, tamarind seed polysaccharide was blended with low methoxy pectin as mucoadhesive excipient and calcium chloride aqueous solutions were used as ionic cross-linking solutions. The impacts of low methoxy pectin and tamarind seed polysaccharide contents in the bead formula were statistically optimized on the encapsulation efficiency and cumulative release (in vitro) of metformin HCl at 10 h employing 3^2 factorial design and response surface methodology. The optimized ionically gelled pectinate-tamarind seed polysaccharide beads of metformin HCl showed $95.12 \pm 4.26\%$ of metformin HCl encapsulation efficiency, $46.53 \pm 3.28\%$ of cumulative release (in vitro) of metformin HCl at 10 h, and 1.93 ± 0.26 mm of mean diameter. In vitro metformin HCl release from these ionically gelled pectinate-based mucoadhesive beads followed controlled-release (zero-order) pattern with super case-II transport mechanism over a period of 10 h (Fig. 11).

The in vitro swelling tests demonstrated that the pH of the test medium greatly influenced the in vitro swelling as well as degradation of these ionically gelled beads.

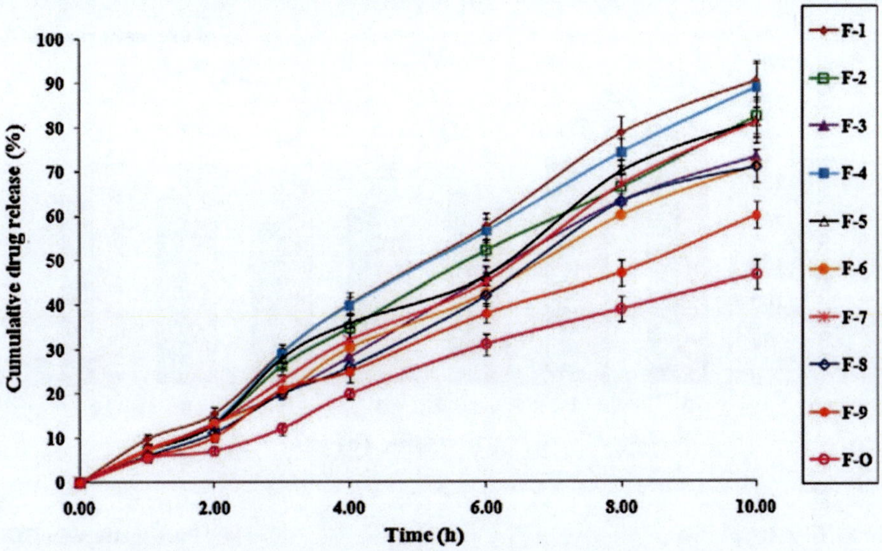

Fig. 11 In vitro drug release from various ionically gelled pectinate-tamarind seed polysaccharide mucoadhesive beads of metformin HCl [Mean ± standard deviation, $n = 3$] [69] (with permission, Copyright © 2013 Elsevier Ltd.)

The optimized ionically gelled pectinate-tamarind seed polysaccharide mucoadhesive beads of metformin HCl also demonstrated excellent *ex vivo* mucoadhesivity onto the excised goat intestinal mucosa and significant *in vivo* hypoglycemic action in the diabetic rats over prolonged period after oral intake (Fig. 12).

Starches extracted from jackfruit (*Artocarpus heterophyllus* Lam.) seeds were blended with ionically gelled pectinate systems to formulate mucoadhesive beads of metformin HCl employing calcium chloride as ionic cross-linker [74]. In this research, the impacts of low methoxy pectin and jackfruit seed starch contents in the bead formula were statistically optimized on the encapsulation efficiency and cumulative release (in vitro) of metformin HCl at 10 h employing 3^2 factorial design and response surface methodology. The optimized ionically gelled pectinate-jackfruit seed starch beads of metformin HCl showed beads of metformin HCl showed 94.11 ± 3.92% of metformin HCl encapsulation efficiency, 48.88 ± 2.02% of cumulative release (in vitro) of metformin HCl at 10 h, and 2.06 ± 0.20 mm of mean diameter. The morphological analysis of the optimized metformin HCl-loaded ionically gelled pectinate-jackfruit seed starch beads was visualized by SEM imaging (Fig. 13), showing an almost spherically shaped bead with free from agglomeration.

These metformin HCl-loaded ionically gelled pectinate-jackfruit seed starch beads demonstrated a sustained controlled releasing of loaded metformin HCl, in vitro, over a period of 10 h (Fig. 14).

The optimized beads also exhibited an in vitro pH-sensitive swelling and excellent *ex vivo* mucoadhesivity onto the excised goat intestinal mucosa over a longer period.

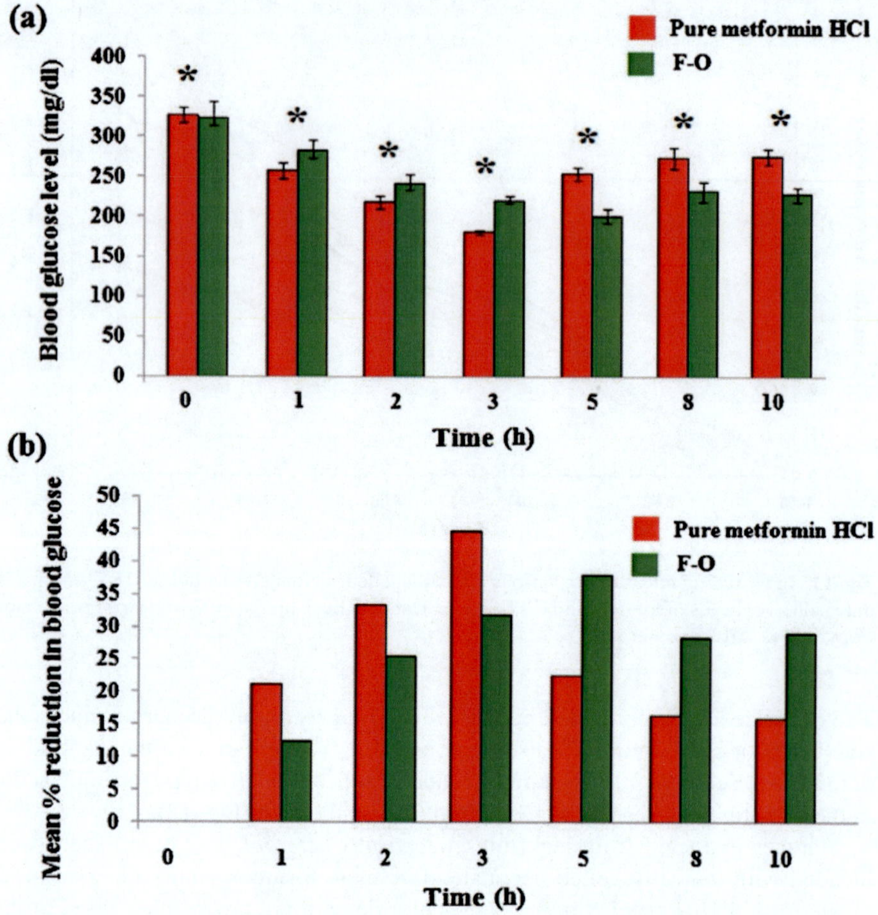

Fig. 12 **a** Comparative *in vivo* blood glucose level in alloxan-induced diabetic rats after oral administration of pure metformin HCl and optimized ionically gelled pectinate-tamarind seed polysaccharide mucoadhesive beads of metformin HCl (F–O) [Mean ± standard deviation, $n = 6$]. The data were analyzed for significant differences ($*p < 0.05$) by paired samples t-test. The statistical analysis was conducted using MedCalc software version 11.6.1.0; **b** Comparative *in vivo* mean percentage reduction in blood glucose level in alloxan-induced diabetic rats after oral administration of pure metformin HCl and optimized ionically gelled pectinate-tamarind seed polysaccharide mucoadhesive beads of metformin HCl (F–O) [69] (with permission, Copyright © 2013 Elsevier Ltd.)

The optimized ionically gelled pectinate-jackfruit seed starch mucoadhesive beads of metformin HCl demonstrated a significant hypoglycemic action the diabetic rats, *in vivo*, over prolonged period after oral intake.

In a research by Sharma and Ahuja [65], thiol modification of pectin was carried out and the synthesized thiolated pectin was used to prepare mucoadhesive beads of metformin HCl *via* ionic gelation employing calcium chloride as ionic cross-linker

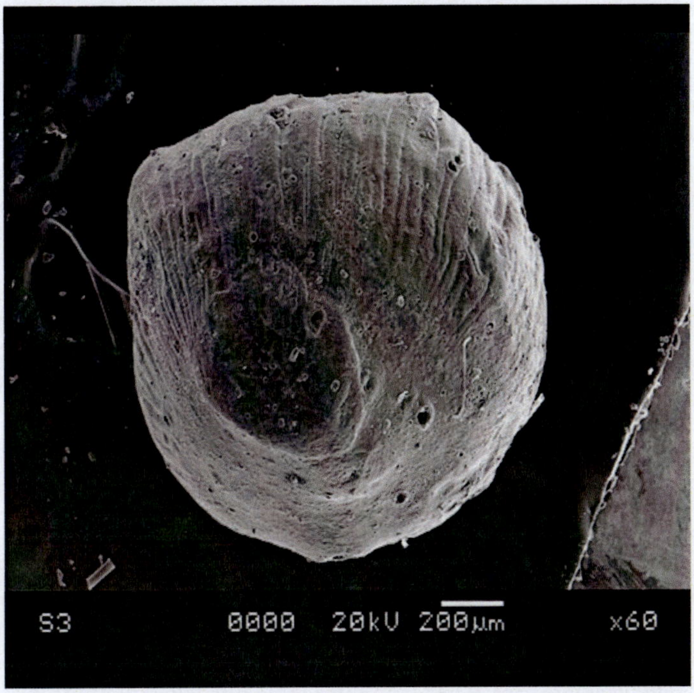

Fig. 13 Scanning electron microphotograph of the optimized metformin HCl-loaded ionically gelled pectinate-jackfruit seed starch beads (F–O) [74] (with permission, Copyright © 2013 Elsevier B.V.)

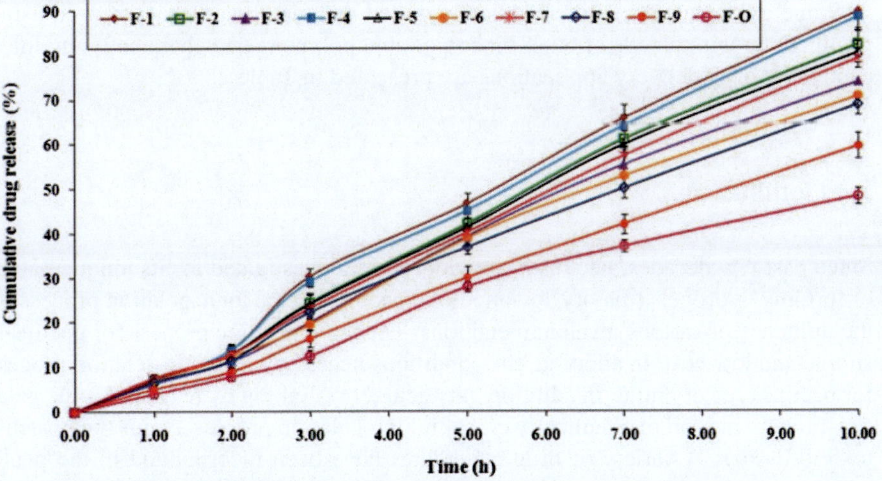

Fig. 14 In vitro drug release from various metformin HCl-loaded ionically gelled pectinate-jackfruit seed starch beads [Mean ± standard deviation, $n = 3$] [74] (with permission, Copyright © 2013 Elsevier B.V.)

Fig. 15 In vitro release profile of metformin from ionically gelled metformin HCl-loaded non-thiolated and thiolated pectinate beads in phosphate buffer (pH 6.8) [65] (with permission, Copyright © 2011 Elsevier Ltd.)

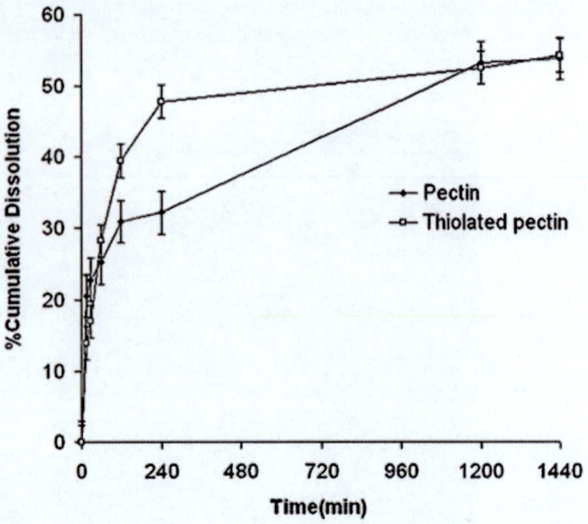

[65]. In this work, the thiolated pectin was chemically synthesized *via* the pectin esterification process using thioglycolic acid and hydrochloric acid. Thiol modification of pectin at 0.6 mmol/g of degree of thiolation did not affect the ionic cross-linking induced by the calcium ions. The formulated ionically gelled metformin HCl-loaded thiolated pectinate beads were also characterized for in vitro drug releasing in phosphate buffer (pH 6.8) and the results indicated almost similar fashion of drug releasing as compared to that of the ionically gelled metformin HCl-loaded non-thiolated pectinate beads (Fig. 15).

Some examples of modified ionically gelled pectinate matrices (prepared using pectin along with blends of other biocompatible polymers and chemically modified pectin) for drug delivery applications are presented in Table 2.

5 Conclusion

Since past few decades, pectins have widely been investigated for its unique nature of forming hydrogels (mainly, low methoxy pectin) *via* the ionic gelation process by the influence of various divalent metal ions. The ionic gelation process for pectins is simple and low cost. In addition, the conditions necessary for ionic gelation process for pectins are very mild. In addition, physical cross-linking by reason of ionic gelation process instead of chemically cross-linking gelation process avoids the possible toxicity issues. A variety of drug candidates have been incorporated in the pectinate beads formed by ionic gelation of low methoxyl pectin for sustained controlled drug delivery applications. All incorporated drugs showed different patterns of drug entrapments and release profiles depending on their method of preparations, compositions, concentration of ionic cross-linking, properties of incorporated drugs, etc.

Table 2 Modified ionically gelled pectinate matrices (prepared using pectin along with blends of other biocompatible polymers and chemically modified pectin) for drug delivery applications

Modified ionically gelled pectinate matrices	Drugs incorporated	References
Alginate/pectin microgels	Antacid (magnesium hydroxide)	Chen et al. [103]
Calcium alginate-pectinate bioadhesive microspheres	Aceclofenac	Chakraborty et al. [104]
Chitosan-calcium pectinate beads	Theophylline and prednisolone	Mennini et al. [105]
Chitosan-pectin composite particles	Bovine serum albumin	Chang and Lin [106]
Chitosan–pectin multiparticulate systems associated with enteric polymers like hydroxypropylmethyl cellulose phthalate and cellulose acetate phthalate	Triamcinolone	Oliveira et al. [107]
Calcium oligochitosan-pectin microcarriers	Fluorescently labeled dextran	Stealey et al. [97]
Calcium pectinate-polyvinyl pyrrolidone beads	Aceclofenac	Nayak et al. [71]
Calcium pectinate-fenugreek seed mucilage mucoadhesive beads	Metformin HCl	Nayak et al. [99]
Calcium pectinate-ispaghula husk mucilage mucoadhesive beads	Metformin HCl	Nayak et al. [58]
Zinc pectinate-ispaghula husk mucilage beads	Aceclofenac	Guru et al. [98]
Linum seed mucilage and hibiscus leaf gum-pectinate mucoadhesive beads	Capecitabine	Kurra et al. [108]
High-amylose corn starch/pectin blend microparticles and tablets	Diclofenac sodium	Desai [109, 110]
Amidated pectin hydrogel beads	Chloroquine	Munjeri et al. [95]
Amidated pectin hydrogel beads	Insulin	Musabayane et al. [111]
Amidated pectin microparticles through Eudragit S 100 coated capsule	Sulfasalazine	Deshmukh et al. [112]
Zinc pectinate-bora rice microspheres	Glipizide	Ramteke and Nath [113]
Zinc pectinate microparticles reinforced with chitosan	Progesterone	Gadalla et al. [114]
Floating beads made of pectin, alginate gelatin, and hydroxypropyl methylcellulose	Gliclazide	Awasthi and Kulkarni [115]

(continued)

Table 2 (continued)

Modified ionically gelled pectinate matrices	Drugs incorporated	References
Pectin/carboxymethyl tamarind seed polysaccharide beads for wound healing	Moxifloxacin	Pandit et al. [116]
Ionically cross-linked films of sodium alginate and pectin for wound dressing	Cefazolin	Shahzad et al. [117]
Mucoadhesive beads of gellan gum/pectin	Ketoprofen	Prezotti et al. [94]
Ionically gelled pectin/modified nano-carbon sphere nanocomposite gel films	5-Fluorouracil	Wang et al. [118]
Calcium pectinate-silica gel beads	Mesalazine	Günter et al. [96]

Though ionically gelled pectinate hydrogel matrices have proved their capability as drug delivery matrices, some disadvantages like low entrapment efficiency and premature release of incorporated small molecular drugs presented by the solubility and swellability of ionically gelled pectinate hydrogel matrices in the acidic milieu. To overcome these drawbacks, modifications of ionically gelled pectinate hydrogel matrices by using biocompatible polymer-blends with low methoxy pectin and chemically modified pectins (like amidated pectins, thiolated pectin, etc.) have been accomplished to prevent premature release of incorporated drugs.

References

1. Taghipour YD, Hokmabad VR, Del Bakhshayesh AR, Asadi N, Salehi R, Nasrabadi HT (2020) The application of hydrogels based on natural polymers for tissue engineering. Curr Med Chem 27(16):2658–2680
2. George A, Shah PA, Shrivastav PS (2019) Natural biodegradable polymers based nano-formulations for drug delivery: A review. Int J Pharm 561:244–264
3. Nayak AK, Hasnain MS, Pal K, Banerjee I, Pal D (2020) Gum-based hydrogels in drug delivery. In: Pal K, Banerjee I, Sarkar P, Kim D, Deng W-P, Dubey NK, Majumder K (eds), Biopolymer-based formulations, biomedical and food applications. Elsevier, United States, pp 605–645
4. Nayak AK, Pal D (2017) Natural starches-blended ionotropically-gelled micrparticles/beads for sustained drug release, In: Thakur VK, Thakur MK, Kessler MR (eds) Handbook of composites from renewable materials, volume 8, nanocomposites: advanced applications, Wiley-Scrivener, USA, pp 527–560
5. Hasnain MS, Ahmed SA, Alkahtani S, Milivojevic M, Kandar CC, Dhara AK, Nayak AK (2020) Biopolymers for Drug Delivery. In: Nayak AK, Hasnain MS (eds) Advanced biopolymeric systems for drug delivery. Springer, Switzerland, pp 1–29
6. Pal D, Nayak AK (2015) Interpenetrating polymer networks (IPNs): Natural polymeric blends for drug delivery, In: Mishra M (ed) Encyclopedia of biomedical polymers and polymeric biomaterials. Taylor and Francis, New York, Vol. VI, pp 4120–4130

7. Hasnain MS, Ahmad SA, Chaudhary N, Hoda MN, Nayak AK (2019) Biodegradable polymer matrix nanocomposites for bone tissue engineering, In: Inamuddin, Asiri AM, Mohammad A (eds) Applications of nanocomposite materials in orthopedics. Woodhead Publishing Series in Biomaterials, Elsevier, United States, pp 1–37
8. Hasnain MS, Rishishwar P, Rishishwar S, Ali S, Nayak AK (2018) Extraction and characterization of cashew tree (*Anacardium occidentale*) gum; use in aceclofenac dental pastes. Int J Biol Macromol 116:1074–1081
9. Hasnain MS, Rishishwar P, Ali S, Alkahtani S, Tabish M, Milivojevic M, Ansari MT, Nayak AK (2020) Formulation and *ex vivo* skin permeation of lidocaine HCl topical gels using dillenia (*Dillenia indica* L.) fruit gum. Rev Mex Ing Quím 19:1465–1476
10. Hasnain MS, Rishishwar P, Rishishwar S, Ali S, Nayak AK (2018) Isolation and characterization of *Linum usitatisimum* polysaccharide to prepare mucoadhesive beads of diclofenac sodium. Int J Biol Macromol 116:162–172
11. Nayak AK, Pal D, Santra K (2016) Swelling and drug release behavior of metformin HCl-loaded tamarind seed polysaccharide-alginate beads. Int J Biol Macromol 82:1023–1027
12. Hasnain SMM, Hasnain MS, Nayak AK (2019) Natural polysaccharides: Sources and extraction methodologies. In: Hasnain MS, Nayak AK, (eds) Natural polysaccharides in drug delivery and biomedical applications. Academic Press, Elsevier, United States, pp 1–14
13. Nayak AK, Bera H (2019) *In situ* polysaccharide-based gels for topical drug delivery applications. In: Maiti S, Jana S (eds) Polysaccharide carriers for drug delivery. Woodhead Publishing, Elsevier, United States, pp 615–638
14. Nayak AK, Pal D (2012) Natural polysaccharides for drug delivery in tissue engineering. Everyman's Sci XLVI:347–352
15. Nayak AK, Hasnain MS (2019) Plant polysaccharides in drug delivery applications. In: Nayak AK, Hasnain MS (eds) Plant polysaccharides-based multiple-unit systems for oral drug delivery. Springer, Singapore, pp 19–23
16. Nayak AK, Ahmed SA, Tabish M, Hasnain MS (2019) Natural polysaccharides in tissue engineering application. In: Hasnain MS, Nayak AK, (eds) Natural polysaccharides in drug delivery and biomedical applications. Academic Press, Elsevier, United States, pp 531–548
17. Pal D, Saha S, Nayak AK, Hasnain MS (2019) Marine-derived polysaccharides: Pharmaceutical applications. In: Nayak AK, Hasnain MS, Pal D (eds) Natural polymers for pharmaceutical applications, Vol. II: Marine and microbiologically derived polymers. Apple Academic Press, ppssss 1–36
18. Nayak AK, Ansari MT, Pal D, Hasnain MS (2019) Hyaluronic acid (hyaluronan): Pharmaceutical applications. In: Nayak AK, Hasnain MS, Pal D, (eds) Natural polymers for pharmaceutical applications, Vol. III: animal derived polymers. Apple Academic Press, USA, pp 1–32
19. Nayak AK, Pal D, Santra K (2015) Screening of polysaccharides from tamarind, fenugreek and jackfruit seeds as pharmaceutical excipients. Int J Biol Macromol 79:756–760
20. Nayak AK, Pal D (2017) Tamarind seed polysaccharide: An emerging excipient for pharmaceutical use. Ind J Pharm Edu Res 51:S136–S146
21. Nayak AK, Bera H, Hasnain MS, Pal D. (2018) Synthesis and characterization of graft-copolymers of plant polysaccharides, In: Thakur VK (ed) Biopolymer grafting, synthesis and properties. Elsevier, The Netherlands, pp 1–62
22. Barbosa MA, Granja PL, Barrias CC, Amaral IF (2005) Polysaccharides as scaffolds for bone regeneration. ITBM-RBM. 26:212–217
23. Nayak AK, Pal D, J. Pradhan, T. Ghorai (2012) The potential of *Trigonella foenum-graecum* L. seed mucilage as suspending agent. Ind J Pharm Edu Res 46:312–317
24. Nayak AK, Ara TJ, Hasnain MS, Hoda N (2018) Okra gum-alginate composites for controlled releasing drug delivery, In: Inamuddin, Asiri AM, Mohammad A (eds), Applications of nanocomposite materials in drug delivery. Woodhead Publishing Series in Biomaterials, Elsevier, United States, pp 761–785
25. Nayak AK, Pal D, Pradhan J, Hasnain MS (2013) Fenugreek seed mucilage-alginate mucoadhesive beads of metformin HCl: Design, optimization and evaluation. Int J Biol Macromol 54:144–154

26. Racovita S, Vasilu S, Popa M, Luca C (2009) Polysaccharides based micro- and nanoparticles obtained by ionic gelation and their applications as drug delivery systems. Rev Rom de Chim 54:709–718
27. Nayak AK, Pal D (2016) Plant-derived polymers: Ionically gelled sustained drug release systems, In: Mishra M (ed.) Encyclopedia of biomedical polymers and polymeric biomaterials. Taylor and Francis: New York, Vol. VIII, pp 6002–6017
28. Reddy K, Mohan GK, Satla S, Gaikwad S (2011) Natural polysaccharides: versatile excipients for controlled drug delivery systems. Asian J Pharm Sci 6:275–286
29. Patil JS, Kamalapur MV, Marapur SC, Kadam DV (2010) Ionotropic gelation and polyelectrolyte complexation: the novel techniques to design hydrogel particulate sustained, modulated drug delivery system: A review. Digest J Nanomater Biostr 5:241–248
30. Nayak AK, Hasnain MS (2020) Ionotropically gelled alginate particles in sustained drug release. In: Nayak AK, Hasnain MS (eds) Alginates in drug delivery. Elsevier, Academic Press, United States, pp 203–230
31. Mohanta BC, Javed MN, Hasnain MS, Nayak AK (2020) Polyelectrolyte complexes of alginate for controlling drug release. In: Nayak AK, Hasnain MS (eds) Alginates in drug delivery. Elsevier, Academic Press, United States, pp 297–321
32. Ray P, Maity M, Barik H, Sahoo GS, Hasnain MS, Hoda MN, Nayak AK (2020) Alginate-based hydrogels for drug delivery applications. In: Nayak AK, Hasnain MS (eds) Alginates in drug delivery. Elsevier, Academic Press, United States, pp 41–70
33. Ali SA, Nayak AK, Banerjee S, Sen KK, Prabhakar T (2020) *In silico* molecular docking of vetiver oil and formulation of vetiver oil-encapsulated gellan gum-based microcapsules for antidepressant activity. Res J Pharm Technol 13:3135–3142
34. Ali SA, Nayak AK, Sen KK, Prabhakar T (2019) Preparation and characterization of vetiver oil encapsulated polymeric microcapsules for sedative and hypnotic activity. Int J Res Pharm Sci 10(4):3616–3625
35. Sinha P, Ubaidulla U, Nayak AK (2015) Okra (*Hibiscus esculentus*) gum-alginate blend mucoadhesive beads for controlled glibenclamide release. Int J Biol Macromol 72:1069–1075
36. Das B, Dutta S, Nayak AK, Nanda U (2014) Zinc alginate-carboxymethyl cashew gum microbeads for prolonged drug release: Development and optimization. Int J Biol Macromol 70:505–515
37. Nayak AK, Khatua S, Hasnain MS, Sen KK (2011) Development of alginate-PVP K 30 microbeads for controlled diclofenac sodium delivery using central composite design. DARU J Pharm Sci 19(5):356–366
38. Pal D, Nayak AK (2011) Development, optimization and anti-diabetic activity of gliclazide-loaded alginate-methyl cellulose mucoadhesive microcapsules. AAPS PharmSciTech 12(4):1431–1441
39. Nayak AK, Pal D (2011) Development of pH-sensitive tamarind seed polysaccharide-alginate composite beads for controlled diclofenac sodium delivery using response surface methodology. Int J Biol Macromol 49:784–793
40. Pal D, Nayak AK (2012) Novel tamarind seed polysaccharide-alginate mucoadhesive microspheres for oral gliclazide delivery. Drug Deliv 19:123–131
41. Nayak AK, Das B, Maji R (2012) Calcium alginate/gum Arabic beads containing glibenclamide: Development and in vitro characterization. Int J Biol Macromol 51:1070–1078
42. Malakar J, Nayak AK, Das A (2013) Modified starch (cationized)-alginate beads containing aceclofenac: Formulation optimization using central composite design. Starch—Stärke 65:603–612
43. Malakar J, Nayak AK, Jana P, Pal D (2013) Potato starch-blended alginate beads for prolonged release of tolbutamide: Development by statistical optimization and in vitro characterization. Asian J Pharm 7:43–51
44. Nayak AK, Pal D, Hasnain MS (2013) Development, optimization and in vitro-in vivo evaluation of pioglitazone-loaded jackfruit seed starch-alginate beads. Current Drug Deliv 10:608–619

45. Nayak AK, Pal D (2013) Ionotropically-gelled mucoadhesive beads for oral metformin HCl delivery: Formulation, optimization and antidiabetic evaluation. J Sci Indus Res 72:15–22
46. Nayak AK, Pal D (2013) Formulation optimization of jackfruit seed starch-alginate mucoadhesive beads of metformin HCl. Int J Biol Macromol 59:264–272
47. Malakar J, Das K, Nayak AK (2014) In *situ cross-linked matrix tablets for sustained salbutamol sulfate release—formulation development* by *statistical optimization*. Polym Med 44:221–230
48. Nayak AK, Pal D, Santra K (2014) Tamarind seed polysaccharide-gellan mucoadhesive beads for controlled release of metformin HCl. Carbohydr Polym 103:154–163
49. Nayak AK, Pal D, Santra K (2014) *Artocarpus heterophyllus* L. seed starch-blended gellan gum mucoadhesive beads of metformin HCl. Int J Biol Macromol 65:329–339
50. Nayak AK, Pal D (2014) *Trigonella foenum-graecum* L. seed mucilage-gellan mucoadhesive beads for controlled release of metformin HCl. Carbohydr Polym 107:31–40
51. Nayak AK, Pal D, Santra K (2014) Ispaghula mucilage-gellan mucoadhesive beads of metformin HCl: Development by response surface methodology. Carbohydr Polym 107:41–50
52. Sinha P, Ubaidulla U, Hasnain MS, Nayak AK, Rama B (2015) Alginate-okra gum blend beads of diclofenac sodium from aqueous template using $ZnSO_4$ as a cross-linker. Int J Biol Macromol 79:555–563
53. Hagesaether E, Bye R, Sande SA (2008) *Ex vivo* mucoadhesion of different zinc-pectinate hydrogel beads. Int J Pharm 347(1–2):9–15
54. Das S, Ng K-Y (2010) Resveratrol-loaded calcium pectinate beads: Effects of formulation parameters on drug release characteristics. J Pharm Sci 99:840–860
55. Das S, Ng K-Y, Ho PC (2010) Formulation and optimization of zinc-pectinate beads for the controlled delivery of resveratrol. AAPS PharmSciTech 2010(11):729–742
56. Nayak AK, Beg S, Hasnain MS, Malakar J, Pal D (2018) Soluble starch-blended Ca^{2+}-Zn^{2+}-alginate composites-based microparticles of aceclofenac: Formulation development and in vitro characterization. Fut J Pharm Sci 4:63–70
57. Patil P, Chavanke D, Wagh M (2012) A review on ionotropic gelation method: Novel approach for controlled gastroretentive gelispheres. Int J Pharm Pharma Sci 4:27–32
58. Nayak AK, Pal D, Santra K (2014) Development of pectinate-ispagula mucilage mucoadhesive beads of metformin HCl by central composite design. Int J Biol Macromol 66:203–221
59. Swamy BY, Yun YS (2015) In vitro release of metformin from iron (III) cross-linked alginate-carboxymethyl cellulose hydrogel beads. Int J Biol Macromol 77:114–119
60. Hosny EA, Al-Helw AA-RM (1998) Effect of coating of aluminium carboxymethylcellulose beads on the release and bioavailability of diclofenac sodium. Pharm Acta Helvet 72:255–261
61. Desai KG (2016) Chitosan nanoparticles prepared by ionotropic gelation: An overview of recent advances. Crit Rev Ther Drug Carrier Syst 33(2):107–158
62. Munarin F, Tanzi MC, Petrini P (2012) Advances in biomedical applications of pectin gels. Int J Biol Macromol 51:681–689
63. Naqash F, Masoodi FA, Rather SA, Wani SM, Gani A (2017) Emerging concepts in the nutraceutical and functional properties of pectin-A Review. Carbohydr Polym 168:227–239
64. Lara-Espinoza C, Carvajal-Millán E, Balandrán-Quintana R, López-Franco Y, Rascón-Chu A (2018) Pectin and pectin-based composite materials: Beyond food texture. Molecules 23(4):942
65. Sharma R, Ahuja M (2011) Thiolated pectin: Synthesis, characterization and evaluation as a mucoadhesive polymer. Carbohydr Polym 85:658–663
66. Assifoui A, Chambin O, Cayot P (2011) Drug release from calcium and zinc pectinate beads: Impact of dissolution medium composition. Carbohydr Polym 85:388–393
67. Lee J-S, Kim E-J, Chung D, Lee HG (2009) Characteristics and antioxidant activity of catechin-loaded calcium pectinate gel beads prepared by internal gelation. Colloids Surf B: Biointerf 74:17–22
68. Monsoor MA, Kalapathy U, Proctor A (2001) Improved method for determination of pectin degree of esterification by diffuse reflectance Fourier transform infrared spectroscopy. J Agri Food Chem 49:2756–2760

69. Nayak AK, Pal D, Santra K (2014) Development of calcium pectinate-tamarind seed polysaccharide mucoadhesive beads containing metformin HCl. Carbohydr Polym 101:220–230
70. Sriamornsak P, Nunthanid J, Cheewatanakornkool K, Manchun S (2010) Effect of drug loading method on drug content and drug release from calcium pectinate gel beads. AAPS PharmSciTech 11:1315–1319
71. Nayak AK, Kalia S, Hasnain MS (2013) Optimization of aceclofenac-loaded pectinate-poly (vinyl pyrrolidone) beads by response surface methodology. Int J Biol Macromol 62:194–202
72. Axelos MAV, Thibault J-F (1991) Influence of the substituents of the carboxylic groups and of the rhamnose content on the solution properties and flexibility of pectins. Int J Biol Macromol 13:77–82
73. Renard CMGC, Jarvis MC (1999) Acetylation and methylation of homogalacturonans 2: effect on ion-binding properties and conformations. Carbohydr Polym 39:209–216
74. Nayak AK, Pal D (2013) Blends of jackfruit seed starch-pectin in the development of mucoadhesive beads containing metformin HCl. Int J Biol Macromol 62:137–145
75. Jantrawut P, Assifaoui A, Chambin O (2013) Influence of low methoxyl pectin gel textures and in vitro release of rutin from calcium pectinate beads. Carbohydr Polym 97(2):335–342
76. Ghibaudo F, Gerbino E, Hugo AA, Simões MG, Alves P, Costa BFO, Campo Dall´ Orto V, Gómez-Zavaglia A, Simões PN (2018) Development and characterization of iron-pectin beads as a novel system for iron delivery to intestinal cells. Colloids Surf B; Biointerf 170:538–543
77. Chaurasia M, Chourasia MK, Jain NK, Jain A, Soni V, Gupta Y, Jain SK (2008) Methotrexate bearing calcium pectinate microspheres: a platform to achieve colon-specific drug release. Curr Drug Deliv 5(3):215–219
78. Dhalleine C, Assifaouia A, Moularic B, Pellequerc Y, Cayotb P, Lamprechtc A, Chambina O (2011) Zinc-pectinate beads as an in vivo self-assembling system for pulsatile drug delivery. Int J Pharm 414:28–34
79. Chung JT, Zhang Z (2003) Mechanical characterization of calcium pectinate hydrogel for controlled drug delivery. Chem Ind 2003(57):611–616
80. Lee J-S, Chung D, Lee HG (2008) Optimization of calcium pectinate gel beads for sustained-release of catechin using response surface methodology. Int J Biol Macromol 42:340–347
81. Pawar AP, Gadhe AR, Venkatachalam P, Sher P, Mahadik KR (2008) Effect of core and surface cross-linking on the entrapment of metronidazole in pectin beads. Acta Pharm 58(1):78–85
82. El-Gibaly I (2002) Oral delayed-release system based on Zn-pectinate gel (ZPG) microparticles as an alternative carrier to calcium pectinate beads for colonic drug delivery. Int J Pharm 232(1–2):199–211
83. Sriamornsak P (1999) Effect of calcium concentration, hardening agent and drying condition on release characteristics of oral proteins from calcium pectinate gel beads. Eur J Pharm Sci 8:221–227
84. Badve SS, Sher P, Korde A, Pawar AP (2007) Development of hollow/porous calcium pectinate beads for floating-pulsatile drug delivery. Eur J Pharm Biopharm 65:85–93
85. Sriamornsak P, Sungthongjeen S, Puttipipatkhachorn S (2007) Use of pectin as a carrier for intragastric floating drug delivery: Carbonate salt contained beads. Carbohy Polym 67:436–445
86. Atyabi F, Inanloo K, Dinarvand R (2005) Bovine serum albumin-loaded pectinate beads as colonic peptide delivery systems: Preparation and in vitro characterization. Drug Deliv 12:367–375
87. Jain A, Gupta Y, Jain SK (2007) Potential of calcium pectinate beads for target specific drug release to colon. J Drug Target 15(4):285–294
88. Si L, Zhao Y, Huang J, Li S, Zhai X, Li G (2009) Calcium pectinate gel bead intended for oral protein delivery: preparation improvement and formulation development. Chem Pharm Bull (Tokyo) 57(7):663–667
89. Cheng K, Lim LY (2004) Insulin-loaded calcium pectinate nanoparticles: effects of pectin molecular weight and formulation pH. Drug Dev Ind Pharm 30(4):359–367
90. Dupuis G, Chambin O, Génelot C, Champion D, Pourcelot Y (2006) Colonic drug delivery: influence of cross-linking agent on pectin beads properties and role of the shell capsule type. Drug Dev Ind Pharm 32(7):847–855

91. Ansari M, Sadarani B, Majumdar A (2019) Colon targeted beads loaded with pterostilbene: Formulation, optimization, characterization and *in vivo* evaluation. Saudi Pharm J 27(1):71–81
92. Chambin O, Dupuis G, Champion D, Voilley A, Pourcelot Y (2006) Colon-specific drug delivery: Influence of solution reticulation properties upon pectin beads performance. Int J Pharm 321(1–2):86–93
93. Kawadkar J, Chauhan Meenakshi K, Ram A (2010) Evaluation of potential of Zn-pectinate gel (ZPG) microparticles containing mesalazine for colonic drug delivery. Daru 18(3):211–220
94. Prezotti FG, Cury BS, Evangelista RC (2014) Mucoadhesive beads of gellan gum/pectin intended to controlled delivery of drugs. Carbohydr Polym 113:286–295
95. Munjeri O, Hodza P, Osim EE, Musabayane CT (1998) An investigation into the suitability of amidated pectin hydrogel beads as a delivery matrix for chloroquine. J Pharm Sci 87(8):905–908
96. Günter EA, Markov PA, Melekhin AK, Belozerov VS, Martinson EA, Litvinets SG, Popov SV (2018) Preparation and release characteristics of mesalazine loaded calcium pectin-silica gel beads based on callus cultures pectins for colon-targeted drug delivery. Int J Biol Macromol 120(Pt B):2225–2233
97. Stealey S, Guo X, Majewski R, Dyble A, Lehman K, Wedemeyer M, Steeber DA, Kaltchev MG, Chen J, Zhang W (2020) Calcium-oligochitosan-pectin microcarrier for colonic drug delivery. Pharm Dev Technol 25(2):260–265
98. Guru PR, Bera H, Das M, Hasnain MS, Nayak AK (2018) Aceclofenac-loaded *Plantago ovata* F. husk mucilage-Zn^{+2}-pectinate controlled-release matrices. Starch—Stärke 70:1700136
99. Nayak AK, Pal D, Das S (2013) Calcium pectinate-fenugreek seed mucilage mucoadhesive beads for controlled delivery of metformin HCl. Carbohydr Polym 96:349–357
100. Das S, Chaudhury A, Ng KY (2011) Preparation and evaluation of zinc-pectin-chitosan composite particles for drug delivery to the colon: role of chitosan in modifying in vitro and in vivo drug release. Int J Pharm 406(1–2):11–20
101. Maestrelli F, Cirri M, Mennini N, Bragagni M, Zerrouk N, Mura P (2012) Influence of cross-linking agent type and chitosan content on the performance of pectinate-chitosan beads aimed for colon-specific drug delivery. Drug Dev Ind Pharm 38(9):1142–1151
102. Rezvanian M, Ahmad N, Mohd Amin MC, Ng SF (2017) Optimization, characterization, and in vitro assessment of alginate-pectin ionic cross-linked hydrogel film for wound dressing applications. Int J Biol Macromol 97:131–140
103. Chen F, Zhang Z, Deng Z, Zhang R, Fan G, Ma D, McClements DJ (2018) Controlled-release of antacids from biopolymer microgels under simulated gastric conditions: Impact of bead dimensions, pore size, and alginate/pectin ratio. Food Res Int 106:745–751
104. Chakraborty S, Khandai M, Sharma A, Khanam N, Patra CN, Dinda SC, Sen KK (2010) Preparation, in vitro and in vivo evaluation of algino-pectinate bioadhesive microspheres: An investigation of the effects of polymers using multiple comparison analysis. Acta Pharm 60:255–266
105. Mennini N, Furlanetto S, Maestrelli F, Pinzauti S, Mura P (2008) Response surface methodology in the optimization of chitosan-Ca pectinate bead formulations. Eur J Pharm Sci 35:318–325
106. Chang KLB, Lin J (2000) Swelling behaviour and release of protein from chitosan-pectin composite particles. Carbohydr Polym 43:163–169
107. Oliveira GF, Ferrari PC, Carvalho LQ, Evangelista RC (2010) Chitosan–pectin multiparticulate systems associated with enteric polymers for colonic drug delivery. Carbohydr Polym 82:1004–1009
108. Kurra P, Narra K, Puttugunta SB, Kilaru NB, Mandava BR (2019) Development and optimization of sustained release mucoadhesive composite beads for colon targeting. Int J Biol Macromol 139:320–331
109. Desai KGH (2005) Preparation and characteristics of high-amylose corn starch/pectin blend microparticles: A technical note. AAPS PharmSciTech 6: article 30
110. Desai KGH (2007) Properties of tableted high-amylose corn starch/pectin blend microparticles intended for controlled delivery of diclofenac sodium. J Biomedical Appl 21:217–233

111. Musabayane CT, Munjeri O, Bwititi P, Osim EE (2000) Orally administered, insulin-loaded amidated pectin hydrogel beads sustain plasma concentrations of insulin in streptozotocin-diabetic rats. J Endocrinol 164(1):1–6
112. Deshmukh R, Harwansh RK, Paul SD, Shukla R (2020) Controlled release of sulfasalazine loaded amidated pectin microparticles through Eudragit S 100 coated capsule for management of inflammatory bowel disease. J Drug Deliv Sci Technol 55:
113. Ramteke KH, Nath L (2014) Formulation, evaluation and optimization of pectin-bora rice beads for colon targeted drug delivery system. Adv Pharm Bull 4(2):167–177
114. Gadalla HH, Soliman GM, Mohammed FA, El-Sayed AM (2016) Development and in vitro/in vivo evaluation of Zn-pectinate microparticles reinforced with chitosan for the colonic delivery of progesterone. Drug Deliv 23(7):2541–2554
115. Awasthi R, Kulkarni GT (2012) Development of novel gastroretentive floating particulate drug delivery system of gliclazide. Curr Drug Deliv 9(5):437–451
116. Pandit AP, Koyate KR, Kedar AS, Mute VM (2019) Spongy wound dressing of pectin/carboxymethyl tamarind seed polysaccharide loaded with moxifloxacin beads for effective wound heal. Int J Biol Macromol 140:1106–1115
117. Shahzad A, Khan A, Afzal Z, Umer MF, Khan J, Khan GM (2019) Formulation development and characterization of cefazolin nanoparticles-loaded cross-linked films of sodium alginate and pectin as wound dressings. Int J Biol Macromol 124:255–269
118. Wang SY, Meng YJ, Li J, Liu JP, Liu ZQ, Li DQ (2020) A novel and simple oral colon-specific drug delivery system based on the pectin/modified nano-carbon sphere nanocomposite gel films. Int J Biol Macromol 157:170–176

Chapter 2
Ionically Gelled Alginates in Drug Delivery

Sreejan Manna, Mainak Mal, Sanchita Das, Dipika Mandal, and Manas Bhowmik

Abstract Alginate is a linearly branched natural anionic polysaccharide obtained from brown seaweed (algae). Due to its versatile abilities like water solubility, biocompatibility, biodegradability and capability to form ionotropic gels, this marine biomaterial is widely used in the food, pharmaceutical, agricultural and biomedical industry. Alginates form ionotropic gels by interacting with divalent metal cations like Calcium ions and trivalent metal cations like Aluminium ions. So far alginates have been profoundly used in the development of various drug delivery systems either alone or along with another biopolymer. This book chapter deals with the various applications of ionotropically gelled alginate based systems for different drug delivery systems like in situ gels, beads and alginate based polymeric blends, along with a brief note about its biological sources, extraction process, chemistry and mechanism of ionotropic gelation.

Keywords Alginate · Ionic gelation · Mechanism of gelation · Drug delivery application

---

S. Manna · S. Das
Department of Pharmaceutical Technology, Brainware University, Kolkata, West Bengal 700125, India

M. Mal
Department of Pharmaceutical Technology, NSHM Knowledge Campus, 124 B L Saha Road, Kolkata 700053, India

D. Mandal
Department of Pharmaceutical Technology, University of North Bengal, Darjeeling, West Bengal, India

M. Bhowmik (✉)
Department of Pharmaceutical Technology, Jadavpur University, Kolkata 700032, India
e-mail: mbhowmik.pharmacy@jadavpuruniversity.in

A. K. Nayak et al. (eds.), *Ionically Gelled Biopolysaccharide Based Systems in Drug Delivery*, Gels Horizons: From Science to Smart Materials,
https://doi.org/10.1007/978-981-16-2271-7_2

1 Introduction

The last thirty years have witnessed a remarkable development of drug delivery systems especially in the field of sustained and controlled release systems. Biocompatible synthetic and natural polymers have been extensively used as drug delivery carriers [1]. The current trend involves the usage of natural polymers for drug delivery owing to their negligible toxicity and biodegradability. Of the chemically diverse natural polymers used for preparation of various drug delivery systems polysaccharides occupy a dominant position. From a conventional point of view, polysaccharides have also been popularly used as emulgents, suspending agents, gelling agents, raft forming agents and as binders in tables [2, 3].

Alginate is a linearly branched natural anionic polysaccharide obtained from seaweed. Alginates undergo ionotropic gelation in presence of divalent metal cations like Calcium ions and trivalent metal cations like Aluminium ions. Owing to its advantages like water solubility, biocompatibility, biodegradability and capability to form ionotropic gels, Alginates are widely used in the food, pharmaceutical, agricultural and biomedical industry. So far alginates have been profoundly used in the development of various drug delivery systems either alone or along with another biopolymer [4]. This book chapter deals with the various applications of ionotropically gelled alginate based systems for drug delivery along with a brief note about its source, extraction process, chemistry and mechanism of ionotropic gelation.

2 Alginate

2.1 Sources

The salts of alginic acid and its derivatives are termed alginates [5]. Alginate is the structural components of cell walls of brown micro-algae (Phaeophyceae) [6–8]. It is present as divalent salt of alginic acid in the cell wall and forms the intercellular gel matrix [6]. Alginates are extracted and purified from brown sea weeds (algae). The main brown seaweed species belonging to class Rodophyceae, Ascophyllum, Durvillaea, Ecklonia, Lessonia, Laminaria, Macrocystis and Sargassumare used for commercial alginate production. Although Sargassum is used when other species are not available its alginate is of borderline quality and low yield. The algae mostly utilized for alginate production commercially are*Laminariahyperborean* [9–11], *Macrocystispyrifera* [10, 11], *Laminariadigitata* [9–11], *Ascophyllumnodosum* [12–14] and to a lesser extent from *Laminaria japonica*. All the commercial varieties are listed in Table 1.

Most of the commercial alginates are collected from algae. Alginates from these sources differ in quality and quantity due to geographical variation. But biotechnologically alginates are produced by bacteria species Pseudomonas and Azotobacter. The

Table 1 Biological and geographical sources of alginates

Source	Family	Occurrence	References
Laminariahyperborean	Laminariaceae	Northern Atlantic Ocean	[15]
Laminariadigitata	Laminariaceae	Northern Pacific Ocean and Northern Atlantic Ocean	[15]
Laminaria japonica	Laminariaceae	Seas of China, Japan and Korea	[16]
Laminarialongicruris	Laminariaceae	Island of eastern North America	[17]
Laminariabrasiliensis	Laminariaceae	Northern Pacific Ocean and Northern and Southwestern Atlantic Ocean	[18]
Lessonianigrescens	Laminariaceae	Mahuin of South Chile	[19]
Macrocystispyrifera	Laminariaceae	Pacific coasts of North and South America	[20]
Saccorhizapolyschides	Phyllariaceae	Lower shore of Europe	[21]
Ecklonia maxima	Lessoniaceae	Coast of South Africa	[22]
Eiseniabicyclis	Lessoniaceae	Korea and Japan	[23]
Egregialaevigata	Lessoniaceae	Coastline of western North America	[24]
Elachistafucicola	Chordariaceae	Eastern Pacific Ocean (Canada)	[25]
Ascophyllumnodosum	Fucaceae	Sheltered shores, in sea lochs and west coasts of Ireland and Scotland	[25]
Durvillaeaantarctica	Durvillaeaceae	Coasts of Chile, southern New Zealand, and Macquarie Island	[26]
Dictyosiphonfoeniculaceus	Chordariaceae	North Pacific Ocean: Arctic Ocean, Bering Sea and Aleutian Is., Alaska, to northern Washington; Japan; Russia.	[27]
Himanthaliaelongata	Himanthaliaceae	Baltic Sea, the North Sea and the North-East Atlantic Ocean	[28]
Hormosirabanksii	Hormosiraceae	Littoral zone or in rock pools of Australia and New Zealand	[29]
Scytosiphonlomentaria	Scytosiphonaceae	Littoral zone and favours wave-exposed shores and rock pools Denmark	[30]
Sargassumnatans	Sargassaceae	North Atlantic and costal area of Gujarat, India	[31]
Sargassumfluitans	Sargassaceae	North Atlantic and costal area of Gujarat, India	[32]

high-quality alginate is produced by bacteria as compared to algae. Sources of bacterial alginates are *Azotobactervinelandii* [32], *Pseudomonas aeruginosa* [33], *Pseudomonas putida* [34], *Pseudomonas mendocina* [35], *Pseudomonas fluorescens* [34], and *Pseudomonas syringae* [36]. The *Azotobacter vinelandii* is the best candidate for industrial production of alginate.

2.2 Extraction

Sea weed algae are collected and thoroughly washed with tap water to remove sand and soil. Then follow encrustation and removal of holdfasts. Finally, the prepared algae are washed with deionised water and dried at 60 °C until a constant weight is reached [37, 38]. These are then chopped and ground in hammer mill. The depigmentation of algae and stabilization of alginate is carried out with 2% formaldehyde solution, washing with deionised water and treating with 0.2 M HCl solution for 24 h [39]. Then the algae are washed with deionised water and treated with 2% sodium carbonate solution with continuous stirring up to 5 h [37]. The supernatant fluid is collected employing centrifugation followed by serial washing with ethanol; methanol and acetone resulting in precipitated purified sodium alginate [40]. Finally, sodium alginate is collected and dried. The schematic extraction process of alginate from seaweeds is given below:

Alginate is synthesized and secreted into the extracellular environment by alginate producing bacteria [32–36, 41]. The schematic process is given below:

2.3 Chemistry

Structure elucidation of Alginates was first attempted by Haug et al. in the year 1966. Chemically alginates consist of anionic co-polymers of α-L-guluronic acid unit (G-unit) and β-D-mannuronic acid unit (M-unit) with 1, 4-glycosidic linkages connecting these units. They remain arranged in irregular patterns with varying proportions of GG, MM and MG units (as shown in Fig. 1). This particular arrangement of M and G residues was first discovered by Haug et al. by employing fractionation of alginates and free boundary electrophoresis of hydrolysable products of alginates [42]. However, definite proof of alginate structure was obtained later on using 1H and 13C NMR spectroscopy [43, 44].

Alginates are susceptible to both acidic and alkaline degradation. Rapid degradation of alginate occurs above pH 10.0 and below pH 5.0. The degradation is comparatively more rapid in case of Alginic acid than the Sodium salt form. This is due to intramolecular catalysis by the C-5 carboxyl groups. Therefore Sodium alginate exhibits a better shelf life than Alginic acid when stored in a cool dry place away from sunlight. If kept in the freezer, the shelf life can get extended from several months to several years. Figure 2 illustrates the acid hydrolysis mechanism of alginates [45].

Fig. 1 Representative alginate structure

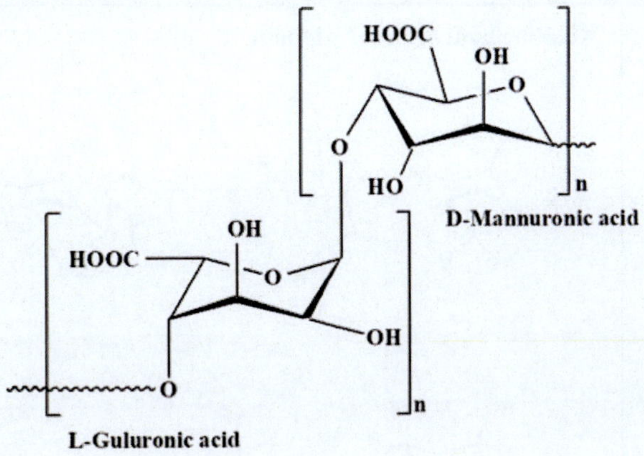

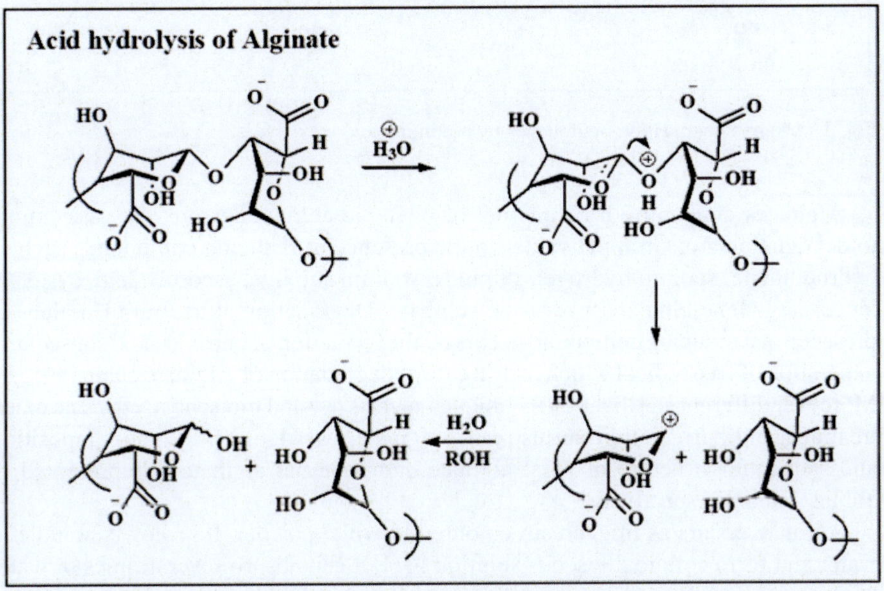

Fig. 2 Acid catalyzed hydrolytic degradation of Alginates

Alginates are also susceptible to enzymatic degradation by lyases. The mechanism of degradation involves a β-elimination and is similar to that of alkaline degradation of alginates. β-elimination occurs by abstraction of the proton at the C-5 position, which is increased by the electron-withdrawing effect of the carbonyl group at C-6 position. Once the C-6 carboxyl group is ionized, the electron-withdrawing effect is reduced and abstraction of the C-5 proton does not occur that rapidly. However, the abstraction occurs at a rate that is still sufficiently rapid resulting in rapid degradation at pH above 10.0. Figure 3 illustrates the alkaline degradation of alginates by β-elimination [46–49].

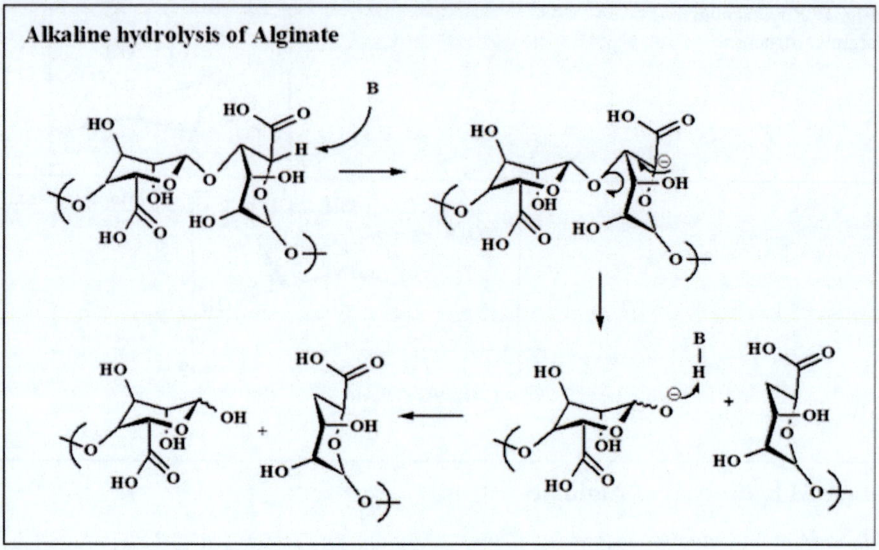

Fig. 3 Alkaline degradation of alginates by β-elimination

Alginates containing high amount of phenolic compounds are also susceptible to degradation at neutral pH values in the presence of reducing compounds such as hydroquinone, sodium hydrogen sulphide, sodium sulphite, ascorbic acid, cysteine, leuco-methylene blue and hydrazine sulphate. Degradation at neutral pH values in presence of a reducing compound occurs by the formation of peroxide and subsequent generation of free radical which in turn causes degradation of Alginate chain [50, 51]. Moreover different sterilization techniques such as dry and moist heat, ethylene oxide treatment and γ-irradiation can also cause alginate degradation [52]. The composition and sequential structure of algal alginate and bacterial alginate is represented in Tables 2 and 3 respectively.

Alginate occurs as fine, cream coloured to white powder. It is slowly soluble in water and form colloidal viscous solution [61]. It can absorb several times of water by its weight (1). It is odourless and tasteless. It is insoluble in ethanol, ethanol/water mixture (ethanol content > 30%), ether, chloroform and organic solvents [62]. It is insoluble in aqueous acidic solution where pH is less than 3. Sodium alginate fibres are insoluble in 60% H_2SO_4 and warm HCl and boiling 5% KOH solution. It is soluble in 5% sodium citrate solution. Various grades of alginates are available in the market having different viscosity. The viscosity may depend on temperature, concentration, pH and presence of metal ions [63–65].

Table 2 Composition and sequential structure of algal alginate

Source	F_G	F_M	F_{MM}	F_{GG}	F_{GM}	F_{MG}	References
Laminariahyperborean-Stipe	0.71	0.29	0.17	0.59	0.12	0.12	[53]
Laminariahyperborean-Leaf	0.51	0.49	0.34	0.36	0.15	0.15	[53]
Laminariadigitata	0.44	0.53	0.47	0.41	0.06	0.06	[40]
Laminaria japonica	0.35	0.65	0.48	0.18	0.17	0.17	[54]
Laminarialongicruris	0.59	0.41	0.07	0.25	0.34	0.034	[55]
Lessonianigrescens	0.38	0.62	0.43	0.19	0.19	0.19	[56]
Macrocystispyrifera	0.39	0.61	0.38	0.16	0.23	0.23	[56]
Ecklonia maxima	0.45	0.55	0.32	0.22	0.32	0.32	[56]
Elachistafucicola	0.78	0.22	0.12	0.68	0.10	0.10	[57]
Ascophyllumnodosum	0.39	0.61	0.35	0.13	0.26	0.26	[58]
Durvillaeaantarctica	0.29	0.71	0.57	0.15	0.14	0.14	[56]
Sargassumfluitans	0.54	0.46	0.36	0.28	0.18	0.18	[37]

Table 3 Composition and sequential structure of bacterial alginate

Source	F_G	F_M	F_{MM}	F_{GG}	% acetylation	References
Azotobactervinelandii	0.08–0.56	0.44–0.92	–	0.03–0.37	11–30	[41]
Pseudomonas aeruginosa	0.23–0.30	0.7–0.77	0.41–0.54	–	0.6–8.8	[59, 60]
Pseudomonas putida	0.22	0.78	0.56	–	18–21	[34]
Pseudomonas mendocina	0.27–0.35	0.65–0.73	0.30–0.46		12–19	[34]
Pseudomonas fluorescens	0.40	0.60	0.20	–	6	[57]

3 Ionic Gelation of Alginate

Alginates exist in nature as alginic acid salts of different metal cations. These metal cations are commonly found in sea water like Na^+, Mg^{2+} and Sr^{2+}. As discussed earlier in the chemistry section, alginates consist of anionic co-polymers of α-L-guluronic acid unit (G-unit) and β-D-mannuronic acid unit (M-unit) with 1, 4-glycosidic linkages connecting these units. They remain arranged in irregular pattern with varying proportions of GG, MM and MG units [66]. The physical characteristics of alginates depend upon the chemistry, sequential arrangement of units and molecular weight [67]. Of the different salts of alginic acid available, the sodium salt of alginic acid i.e. sodium alginate is the most widely used form owing to its solubility in aqueous solvents. Sodium alginate is capable of undergoing ionotropic gelation in diverse aqueous solvents under the influence of divalent cations like Ca^{2+}, Ba^{2+}, Zn^{2+}, etc., and trivalent cations like Al^{3+} [68, 69]. The mechanism of ionotropic gelation involves intermolecular interactions between the carboxyl groups of alginate backbone and these metal ions [70]. The metal cation gets incorporated in the alginate structure's electronegative cavities in such a way that they appear like eggs in an 'Egg Box'. The divalent or trivalent metal cations compete with the Na^+ of sodium alginate in interacting with the carboxyl groups of sodium alginate, thereby bringing the two polyuronate chains together. The cross linking metal ions fit into the interstitial space of the two polyuronate chains [66, 71]. This intermolecular interactions result in the formation of stable reversible cross-linked alginate-gels under ambient formulation conditions. The quality of gelling depends on the species of metal ion and the control of ion addition. Mg^{2+} and monovalent metal ions are unable to induce ionotropic gel formation [72]. Of the different divalent metal ions available $Ca2^+$ is the most popular due to its safety, availability and price [73]. Alginates having high proportion of G units provide gels of high mechanical resistance and modifiable permeability whereas those having high proportion of M units provide advantage in additional coating [74]. The sol gel transformation of alginate is nearly independent of temperature [75]. The ionotropic gelled alginate systems are mostly spherical in shape. Various factors like ease of preparation, absence of toxic organic solvents and cheap production cost have led to the development of diverse drug delivery systems based on ionotropically gelled alginate like microparticles, nanaoparticles, beads, etc. [76–81]. The encapsulation of drug, its release pattern and swellability of ionotropically gelled alginate system are determined by various factors like quality of sodium alginate, its concentration as well as concentration of cross linking metal ions, proportion of polymer to drug, curing time, pH, etc. [78, 81]. Despite the advantages offered by ionotropically gelled alginate systems, it suffers from two serious disadvantages in terms of drug encapsulation and drug release. Firstly, the long immersion time during gel formation may allow the drug to escape thereby resulting in poor encapsulation. Secondly, the drug release from ionotropically gelled alginate system is sudden as the matrices degrade quite rapidly. To overcome these difficulties attempts have been made to blend another polymer with sodium alginate [82–85].

4 Insights into the Use of Ionically Gelled Alginate-Based Systems for Drug Delivery

4.1 Ionically Gelled Systems Made Up of Alginate as Only Biopolymer

Ionically induced gelation of alginic acid was performed to synthesize nanoparticles for persulfate therapy to manage drug-resistant bacteria. Potassium persulfate was used in the de-polymerization reaction of alginic acid at pH 4 for 2 days. The molecular weight of alginic acid was determined by gel permeation chromatography. Low molecular weight nanoformulations were developed with a mean size of

54 ± 0.41 nm. The tetracycline loaded nanoparticles were investigated for minimum inhibitory concentration by using *E.coli* XL-1 which is resistant to tetracycline [86].

Sriamornsak et al. has successfully incorporated theophylline into alginate based hard capsules to investigate the gel-forming ability and drug release in different pH system. Different grades of alginate were used demonstrating different drug release behaviour depending on the composition and pH of the buffer. The study reports indicated variable drug release kinetics based on the presence of calcium acetate. In the acidic medium diffusion-controlled release following non-Fickian mechanism was observed where as in neutral pH super case II mechanism was observed [87].

Another study involving theophylline was aimed to determine the diffusion coefficient through a polymeric membrane of sodium alginate. Different mathematical models were employed for data fitting of drug diffusion. Sodium alginate was used in three different concentrations for hydrogel-based membrane development. The study data indicated that the drug diffusion is reliable on linear model and not greatly dependant on concentration of sodium alginate [88].

In situ gel of sodium alginate was developed for sustained delivery of theophylline. The release of calcium ion from the aqueous solution of sodium alginate induces gelation in acidic medium. The oral bioavailability was determined by using rats and the reports suggested a 1.3–2 fold increase in bioavailability of theophylline. The mean residence time for all the formulation was found similar [89].

Polyglycidyl methacrylate grafted sodium alginate (PGMA-g-SA) was used for pH-triggered delivery of riboflavin. This hydrogel formulation was compared with calcium alginate beads for entrapment, erosion, swelling and drug release. The PGMA-g-SA based hydrogel showed better drug release profile sustained for 3 days in basic medium and up to 4 days in acidic medium. The other parameters like drug entrapment, swelling and degradation of matrix were found superior in respect to calcium alginate beads [90].

The efficacy of a non-selective beta-blocker cartelol was investigated through ophthalmic solution of alginic acid. The alginic acid solution undergoes in situ gel formation and interacts with mucus to reduce the loss of drug through lacrimal fluid. The in vitro release study indicated a slow release of cartelol. Intraocular pressure was measured in rabbits which showed the efficacy of the drug in reducing ocular pressure up to 8 h. An increased ocular bioavailability was observed for in situ gel of alginic acid in respect to cartelol alone [91].

Another sodium alginate based*in situ* gel was developed for ophthalmic delivery of ciprofloxacin hydrochloride. The presence of Ca^{+2} ions in lachrymal fluid-induced gelation of alginate which increased the contact time. HPMC was used as a viscosity modifier. The pH of the formulations was reported between 6.49 and 6.58. The rheological studies were performed and the reports indicated good sol to gel transition ability for the optimized formulation with desired viscosity [92].

Mandal et al. developed an in situ gel for ophthalmic delivery of moxifloxacin hydrochloride. Sodium alginate was used as a mucoadhesive gel-forming polymer that can undergo gelation in presence of calcium ions. HPMC was used as viscosity enhancer. Buffering agents were used to adjust the pH of the formulations to 6.5. The developed in situ gels showed a sustained drug release up to 10 h. Eye irritation was performed by using albino rabbits and the study reports showed no indication of ocular damage or other undesired clinical symptoms [93].

Alginic acids obtained from different sources were investigated as a potential drug carrier by Draget et al. Hydrogel beads of alginic acid were developed and the swelling property was determined. The swelling property and solubility study showed a dependency on homopolymeric blocks forming intermolecular junction and its tendency to decrease the chemical potential of water. The study result showed increased swelling with increase in temperature and reduced average bead size [94].

4.2 *Ionically Gelled Alginate-Based Systems Prepared Using Alginate and Second Biopolymers*

4.2.1 Chitosan–Alginate System

Being one of the most abundant naturally occurring polysaccharide, chitosan is widely used as a biopolymer in the field of drug delivery science. The presence of amino group in chitosan has imparted versatility, such as cationic behaviour, conjugation with other groups, by binding to reactive sites, etc. [95]. Depending on degree of deacetylation, the physicochemical property of chitosan can vary. The non-toxic, biodegradable and biocompatible nature of chitosan has helped it to gain popularity amongst researchers in pharmaceutical field. Chitosan can be easily developed into gelled system including beads [96, 97].

Sodium alginate–chitosan beads were prepared by ionotropic gelation technique with the help of calcium chloride. The concentrations of the polymers were investigated on particle size and drug entrapment. Bovine serum albumin (BSA) was used as a model drug and the entrapment efficiency was reported as 98.5%. A burst release of the drug was observed in pH 6.8 phosphate buffer solution [98].

Zhang et al. developed Ca^{+2} mediated cross-linked sodium alginate microspheres solidified with chitosan. The gelled microspheres were prepared for oral insulin delivery. The size of the microgels was found to be smaller than the alginate droplets. Various study results demonstrated a sustained release of insulin protecting it from protease digestion. Animal testing involving diabetic rats showed strong hypoglycemic activity for 60 h [99].

pH-sensitive sodium alginate chitosan microparticles were developed by ionic gelation method for delivering an antituberculosis drug—rifampicin. The in vitro drug release study results indicated a pH dependant controlled release of rifampicin from microparticles. 20% of the loaded drug release was reported in acidic medium in 2 h, where as in pH 6.8 phosphate buffer complete drug release was observed [100].

Polymyxin B loaded sodium alginate chitosan gelled microparticles were developed by Coppi et al. by using Ca^{+2} ions. The microparticles were designed to be up taken by Peyer's patches for improving the oral bioavailability of polymyxin B.

Spray drying technology was used to prepare microparticles less than 3 μm size range. Restricted drug release was observed in gastric pH protecting the drug in acidic medium and facilitating a sustained release in intestinal environment [101].

Calcium alginate microparticles were prepared for oral delivery of a nonsteroidal anti-inflammatory drug, (NSAID) naproxen with the help of a secondary biopolymer—chitosan oligosaccharide (COS). The potential of COS in delivering a stable microparticulate system was investigated by Čalija et al. The microscopic study revealed slightly deformed spherical microparticles with rough surface texture. High encapsulation efficiency was reported due to low solubility of drug in the gelling medium. Reduced swelling behaviour and erosion was reported for COS modified calcium alginate microparticles facilitating a prolonged drug release [102].

Another alginate–chitosan–calcium complex was developed by Onishi et al. incorporating lactoferrin in microparticulate system for oral delivery. In vitro dissolution study indicated about 60% of drug release in pH 1.2 gastric fluids in 1 h, followed by controlled release up to 7 h. The anti-inflammatory study was performed using carrageenan-induced edema in rats which showed significant decrease in swelling after oral intake of lactoferrin loaded microparticles [103].

Nifedipine was loaded in alginate–chitosan nanoparticles developed by ionic pregelation and polyelectrolyte complexation. The mean diameter of the nanoparticulate formulation was found within the range of 20–50 nm. pH-responsive release behaviour was observed for the mucoadhesive gelled nanoparticles. In respect to acidic medium, release of nifedipine was found to increase in simulated intestinal fluid. The drug release mechanism from nanoparticles was reported as Fickian diffusion [104].

An antiretroviral drug stavudine was successfully incorporated into an alginate–chitosan beads prepared by ionic gelation. Polyelectrolyte complex formation was observed between negatively charged sodium alginate and positively charged chitosan. A sustained drug release more than 12 h was reported following zero-order release mechanism for the alginate–chitosan complex [105].

4.2.2 Locust Bean Gum–Alginate System

Locust bean gum (LBG) is a biopolysaccharide belonging to galactomannans category [106]. LBG is extracted from the seed endosperm of *Ceratonia siliqua* [107]. The solubility of LBG in cold water is less and hence it needs heating for solubilization. The viscous solution of LBG was reported to be unaffected by temperature and pH [108]. This non-starch polysaccharide has gained significant interest in the field of delivering therapeutics at specific site in a controlled manner.

Ionotropic gelation method was followed to develop interpenetrating polymeric network (IPN) microbeads for treating colorectal cancer. Capecitabine, a prodrug used in the treatment of gastric or colon cancer is encapsulated in the microbeads with the help of a cross-linking agent calcium chloride. The in vitro drug release and in vivo pharmacokinetic profile indicated good oral bioavailability of the drug. Swiss albino rats were used for acute oral toxicity study which demonstrated no toxicity.

Sulforhodamine-B assay was performed on HT-29 cell line, and the result exhibited inhibition of cellular growth [109].

Pawar et al. developed sodium alginate based ionically gelled gastroretentive system by using different galactomannans such as, locust bean gum, guar gum and tora seed gum. Captopril was encapsulated in the calcium chloride cross-linked mucoadhesive beads. The change in the ratios of galactomannans significantly altered the mechanical properties of the beads to sustain the drug release. The in vitro dissolution study reported a sustained drug release up to 12 h [110].

Jana et al. developed sodium alginate–LBG microspheres by inotropic gelation technique using calcium ion as a cross-linker for sustained release of aceclofenac. The entrapment efficiency was reported between 59 and 93%. The average size range of the microspheres was found to be 406–684 μm. In vitro drug release study revealed sustained drug release up to 8 h in pH 6.8 phosphate buffer. In vivo pharmacodynamic study was performed in carrageenan injected rats which demonstrated a prolonged anti-inflammatory activity [80].

The impact of hydrogelation period on carboxymethyl locust bean gum–sodium alginate beads was investigated by Dey et al. for delivering an anti-diabetic drug–glipizide. Aqueous solution of aluminium chloride was used as gelation medium. With longer exposure in gelation medium, increased mean size and increased drug loss was reported. A faster drug release was observed for minimum of 8 h in simulated intestinal fluid (pH 7.4 phosphate buffer) in respect to simulated gastric fluid (pH 1.2 HCl buffer) [111].

4.2.3 Tamarind Seed Polysaccharide–Alginate System

Tamarind seed polysaccharide (TSP) is a hydrocolloid obtained from the seeds of *Tamarindus indica* Linn [112]. In aqueous solution, TSP exhibits good gelling property and stability against heat, mechanical impact and acidic environment. Depending on these characteristics, TSP has found many applications as gelling agent, stabilizer and food thickener [113]. The mucoadhesive property as well as matrix forming ability of TSP has made it a suitable biopolymer in drug delivery system.

TSP—alginate microspheres were developed by ionotropic gelation technique for sustained delivery of a potassium channel blocker, dalfampridine. The effect of polymeric ratio, agitation speed and cross-linker concentration was investigated. The process variables were reported to affect the drug entrapment, drug release and mean size of the microspheres. The drug release was found to be sustained over 12 h. The drug release mechanism followed first-order kinetics with non-Fickian diffusion mechanism [114].

Pal and Nayak have developed mucoadhesive microspheres based on TSP—alginate system by varying the ratio of $CaCl_2$ for delivery of an anti-diabetic drug gliclazide. The mean particle diameter was reported between 752 and 948 μm. A prolonged release of gliclazide was observed over 12 h. Aloxan-induced diabetic rats were used to perform in vivo drug release study. The study results indicated significant decrease in blood glucose level in albino rat model [115].

O-carboxymethylated tamarind seed polysaccharide was combined with sodium alginate to form a hydrogel network to incorporate acyclovir by Jana et al. Hydrogel based microparticles were developed by ionic gelation method with the help of a cross-linker. The drug release was found to be controlled by swelling of the polymeric chains. The release of drug from microparticulate system followed Korsmeyer-Peppas kinetics [116].

Bera et al. developed calcium alginate–tamarind gum based floating beads for stomach-specific delivery of risperidone. Magnesium stearate was incorporated during ionotropic gelation method to maintain the low density of the beads. The alginate gel coated polymeric beads have demonstrated a slower release of drugs in stomach. The release of risperidone from floating beads followed Higuchi kinetics with Fickian and anomalous diffusion mechanism [117].

Metformin HCl was successfully entrapped in alginate–TSP beads by ionic gelation technique. The encapsulation efficiency was reported as $94.86 \pm 3.92\%$. The study findings suggested pH dependant swelling and degradation of polymeric beads. A slower drug release was observed in simulated gastric fluid in respect to simulated intestinal fluid. The ionically gelled beads exhibited a sustained drug release following zero-order kinetics with super case II transport [118].

4.2.4 Gum Arabic–Alginate System

Gum Arabic (GA) are dried, edible exudates obtained from the stems of *A. Senegal* (L.) [119]. This complex polysaccharide is either slightly acidic or neutral in nature found as polysaccharidic acid mixed with magnesium, potassium and calcium salts. The major use of GA is in pharmaceutical and cosmetics industries as well as in food industry is as thickening agent, stabilizer and emulsifier [120]. It has gained popularity amongst researchers due to its higher solubility, non-toxicity, pH stability and excellent gelling ability [121].

Calcium alginate–GA beads were developed by Nayak et al. to achieve sustained release of glibenclamide by ionic gelation method. The encapsulation efficiency was reported as $86.02 \pm 2.97\%$. The average size of the beads was found within the range of 1.15–1.55 mm. The in vitro drug release study demonstrated retarded release of $35.68 \pm 1.38\%$ after 7 h. The swelling behaviour of beads was found to be affected by pH of the test medium [71].

Li et al. have investigated the efficacy of extract of *Perinereisa ibuhitensis* by incorporating it into alginate-GA-gelatin based microcapsules. Scanning electron microscopy (SEM) revealed the encapsulation of microcapsules in calcium alginate hydrogel matrix. The in vivo antioxidant activity of *Perinereisa ibuhitensis* extract was evaluated using mice model. The study result indicated a higher O_2^- scavenging than *Perinereisa ibuhitensis* extract. A localized concentration of the extract was observed in small intestine enhancing the absorption and in vivo effectiveness [122].

Liquid core hydrogel beads (LHB) containing sodium alginate and gum arabic solution incorporating total phenolic compounds were developed to prevent its degradation in simulated GI fluid as well as during storage. The developed LHB were reported to be spherical in shape with a mean diameter of 4.63–5.66 mm. A 25% gum arabic in formulation showed maximum hardness while 50% gum arabic showed highest drug loading %. The rate of decay during storage was reported as 6.10×10^{-3} day^{-1} with a $t_{½}$ of 113.63 days. The drug release rate was found very slow ($k = 2.25 \times 10^{-6}$) following Fickian diffusion [123].

Ionotropic gelation technique was employed to develop alginate–gum arabic based beads for controlled release of bovine serum albumin. With the increase in the polymer concentration, the encapsulation efficiency was reported to increase. A pH dependant swelling of the beads retarded the release of BSA in acidic medium and facilitated the release in intestinal fluid. The study report suggested a decreased release of protein with increased concentration of alginate and gum arabic [124].

4.2.5 Cellulose Derivatives–Alginate System

Being one of the most abundant naturally occurring materials, cellulose is extensively used in pharmaceutical industry. It consists of a long chain of anhydro-D-glucopyranose covalently joined together between C1 and C4 carbon atom [125]. The inherent properties of cellulose allow it to modify through etherification, esterification, radical grafting and de-polymerization [126]. The common derivatives of cellulose are hydroxyl propyl cellulose, hydroxyl propyl methyl cellulose, ethyl cellulose, methyl cellulose, sodium carboxy methyl cellulose, etc. [127].

Carboxymethyl cellulose–alginate beads were developed by Kim et al. for oral delivery of protein. Fe^{3+} ions were used as a cross-linker. SEM study revealed the rough and porous surface of the beads observed to be increased with the increase in carboxymethyl cellulose concentration. The swelling behaviour and drug release were studied in different buffer system and the data indicated a controlled drug release over 24 h [128].

Olfactory ensheathing cells were encapsulated in alginate–methyl cellulose hydrogel based delivery system for treating spinal cord injuries. Gelled beads were developed by using $CaCl_2$ as a cross-linking agent. The encapsulation efficiency, immunochemistry and histology study results were found satisfactory. 1:2 alginate–methyl cellulose ratio showed highest drug release as well as maximum cell proliferation [129].

Sharma et al. developed an in situ gelling system based on sodium alginate and methyl cellulose for sustained delivery of paracetamol. 1.5% w/v aqueous dispersion of sodium alginate in presence of calcium ions were resulted in gelled matrix formation. The study revealed that the retention of drug depends on the percentage of polyethylene glycol, methyl cellulose and sodium chloride. The in vitro drug release study showed a diffusion-controlled drug release from the gelled matrix. The oral bioavailability study was performed on rabbit which demonstrated a more sustained release of drug in comparison to marketed suspension of paracetamol with equivalent dose [130].

Another cellulose derivative–sodium Carboxymethyl cellulose was employed to prepare microbeads in combination with sodium alginate for delivering amoxicillin.

Chitosan-coated magnesium aluminium silicate (MAS) was also used to control the release of amoxicillin. The average size of microbeads was reported between 745 and 889 μm. SEM study revealed spherical shape with rough surfaces. The swelling behaviour was studied in distilled water and was found to alter, depending on the ratio of magnesium aluminium silicate. The in vitro drug release indicated the release of drug was dependent on polymeric concentration, ratio of MAS and extent of coating [131].

Sodium alginate was used as a raft forming system in combination with HPMC to achieve sustained release of tinidazole. The efficacy of alginate based raft forming system was evaluated by measuring raft strength and raft volume which was found satisfactory. After completion of raft formation the raft weight was found within the range of 5.21 ± 1.17–7.88 ± 1.95 g. The dissolution study revealed a sustained drug release up to 8 h by following Korsmeyer-Peppas mechanism for majority of the formulations [3].

4.2.6 Sterculia Gum–Alginate System

Sterculia gum or gum karayais a partially acetylated polysaccharide commonly used in food and pharmaceutical industries [132]. This non-toxic, nonallergic and nonteratogenic biopolymer is widely used as a drug carrier. Apart from drug delivery it is also being investigated for wound dressing due to its swelling property, high water retention ability and excellent antimicrobial potential [133]. The most common application of sterculia gum is found in gastrointestinal delivery of therapeutics and in hydrogel based drug delivery [134, 135].

Sterculia gum–alginate based mucoadhesive beads were developed by ionic gelation technique for gastroretentive delivery of resperidone. The percentage of drug release after 8 h was reported as 70.84%. The mucoadhesive floating beads demonstrated excellent buoyancy with significant mucoadhesion property which facilitated a slow release of drug following Korsmeyer-Peppas kinetics [132].

Emulsion gelation technique was employed by Guru et al. to develop aceclofenac loaded alginate–sterculia gum based oil entrapped floating beads. The entrapment efficiency of aceclofenac was reported as $90.92 \pm 2.34\%$. The floating time was found over 8 h facilitating sustained release of aceclofenac in gastric pH. $41.65 \pm 3.97\%$ of drug was reported to release after 7 h of dissolution following Korsmeyer–Peppas kinetics with non-Fickian diffusion mechanism [136].

Pantoprazole was encapsulated in ionically gelled beads of alginate–sterculia gum cross-linked by $CaCl_2$. The synthesized floating beads were evaluated for swelling studies and in vitro drug release. The drug release was found higher in pH 2.2 buffer in comparison to drug released in distilled water. The release of the drug from the core of the gelled beads was found to follow Fickian diffusion mechanism [137].

4.3 Ionically Gelled Alginate-Based Systems Prepared Using Alginate and Inorganic Materials

4.3.1 Calcium–Alginate System

Acyclovir loaded alginate microspheres were synthesized by using $CaCl_2$ as a cross-linker. 66.42 ± 1.01% of maximum mucoadhesion was reported which indicated good mucoadhesive property. The in vitro release profile was investigated in simulated gastric medium resulting 72.12% drug release over 8 h. In vivo Gamma scintigraphy analysis demonstrated a gastroretention over 4 h [138].

Ionically interacted hydrogel microparticles were developed by using sodium alginate and calcium chloride to achieve sustained release of antitubercular drugs—rifampicin, pyrazinamide and isoniazid. The in vitro drug release showed a sustained release of drugs over 20 days in pH 7.4 phosphate buffer. In vivo study indicated a sustained drug release maintaining a plasma concentration for 3–5 days and in organs up to 9 days. A nine fold increase in bioavailability was observed in respect to free drugs [139].

Malakar et al. developed an alginate based floating system by emulsion—gelation technique for gastroretentive delivery of cloxacillin. The optimized formulation showed an entrapment efficiency of 64.63 ± 0.78 with a density of 0.90 ± 0.05 g/cc. A floating lag time of 8.45 min was reported with a floating time over 12 h. Sustained release of cloxacillin was reported over 8 h following Korsmeyer-Peppas kinetics with non-Fickian diffusion mechanism [140].

Bonilla et al. developed highly concentrated O/W and W/O emulsion consisting an aqueous solution of sodium alginate and castor oil with a cross-linker $CaCl_2$. The mean size of droplet was reported in nano range. Clindamycin hydrochloride or ketoprofen were solubilized in concreted emulsion system and the drug release based on the aqueous phase composition was investigated which was found to be controlled over 24 h [141].

Calcium carbonate was used as a cross-linker in the preparation of alginate based floating in situ gel of a proton pump inhibitor, pantoprazole developed by Ramana et al. The viscosity of the prepared gel was found to increase with increasing concentration of sodium alginate. Excellent gelling capacity was observed for the optimized formulation. The in vitro drug release study indicated a sustained release of drug up to 12 h [142].

Porous hydrogel microparticles based on sodium alginate and calcium carbonate were fabricated by Wang et al. for oral delivery of ibuprofen. The strength of the hydrogel was optimized by varying the ratio of alginate and $CaCO_3$. In vitro dissolution study results indicated a porous $CaCO_3$ controlled drug release profile similar to pure ibuprofen [143].

Sodium alginate–calcium bicarbonate based in situ gelling system was developed to deliver baclofen—a skeletal muscle relaxant. Depending on the concentration of sodium alginate and sodium bicarbonate the viscosity, buoyancy time and in vitro drug release were investigated. The release of the drug followed Higuchi model indicating a diffusion controlled release of baclofen [144].

4.3.2 Iron–Alginate System

Amoxicillin loaded iron oxide—alginate hydrogel nanomaterials are synthesized by cross-linking with $FeCl_3$ and $FeCl_2$. The mean size of the developed nanomaterials was reported as 9 nm. The release rate of amoxicillin in simulated gastric fluid was found 11.65% after 4 days and 81.49% in biolysis serum after 3 days. The antibacterial analysis showed significant efficacy against three strains of pathogenic bacteria [145].

Magnetic alginate hydrogel beads were developed with the help of $CoFe_2O_4$ by ionic cross-linking. pH dependant swelling was observed in pH 7.4. The in vitro release behaviour of the beads were found to alter depending on degree of cross-linking and alginate ratio. Cytotoxicity study was performed through MTT assay using U87 cell lines which demonstrated biocompatibility of the delivery system [146].

Iron oxide and gelled alginate nanostructured beads were synthesized through colloidal particles self-assembly at the interface followed by in situ gelation. Insulin microcrystal was incorporated by dispersing with sodium alginate solution during fabrication. The hydrogel core and the inorganic shell provided a double barrier to sustain the drug release. The drug release was found to follow Weibull kinetics maintaining Fickian diffusion [147].

Anhydrous iron chloride and iron sulphate heptahydrate were used to develop magnetically controlled nanocellulose hydrogel beads of alginate for sustained release of ibuprofen. Study report indicated the presence of nanocellulose can enhance the integrity and swelling of hydrogel beads. The in vitro dissolution data indicated a controlled release of ibuprofen which was mathematically fitting to Peppas-Sahlin and Korsmeyer-Peppas model [148].

pH-sensitive magnetic nanoparticles were synthesized by Rashidzadeh et al. for controlled release of diclofenac sodium. Ionotropic gelation technique was used to develop sodium alginate and Fe_3O_4 based hydrogel beads. The dissolution study showed drug release of 83% after 200 min at pH 7.4. The release of diclofenac sodium was also regulated by externally applied magnetic field. The alginate based beads have also demonstrated excellent antibacterial activity [149].

4.3.3 Graphene–Alginate System

Doxorubicin hydrochloride was successfully encapsulated in hydrogel nanocomposites made of grapheme oxide, alginate and chitosan. High drug loading was reported with pH dependant release behaviour of drug. Tumour targeted release was achieved by triggering the drug release in acidic pH. Cellular uptake study indicated internalization of nanocomposites by MCF-7 cells which showed excellent cytotoxicity to tumor cells [150].

Calcium alginate beads were developed by Veerla et al. by varying ratio of graphene oxide. Risedronate was encapsulated in the prepared hydrogel beads. The drug release profile was investigated which showed sustained drug release up to 24 h with increasing concentration of graphene oxide. Cellular investigations demonstrated the effectiveness of microfluidic channel promoting the attachment and proliferation of osteoblast cell [151].

5 Conclusion

The application of ionically gelled alginate in different drug delivery system including in situ gels, beads and alginate based polymeric blends have been summed up in this chapter. We have also outlined the biological sources, extraction, chemical composition and general chemistry of alginate. Being a biomaterial, the efficacy of alginate is already established in various pharmaceutical fields. The biodegradability, biocompatibility and non-toxicity of alginate have aided in its wide application in pharmaceutical research. The ability to form ionotropically gelled mass has added certain advantages to this versatile marine biomaterial. The chemical modification of alginate through ionic gelation to achieve controlled delivery of therapeutics has been highlighted here.

References

1. Aguzzi C, Cerezo P, Viseras C, Carmella S (2007) Use of clay as drug delivery system: possibilities and limitations. Appl Clay Sci 36:22–36
2. Zohurianan-Mehr MJ, Pourjavadi A (2003) New polysaccharide-gPAAm co-polymers: synthesis and thermal characterization. Polym Adv Tech 14:508–516
3. Manna S, Jayasri K, Annapurna KR, Kanthal LK (2017) Alginate based gastro-retentive raft forming tablets for enhanced bioavailability of tinidazole. Int J Appl Pharm 9:16–21
4. Tønnesen HH, Karlsen J (2002) Alginate in drug delivery Systems. Drug Dev Industrial Pharm 28:621–630
5. Szekalska M, Pucibowska A, Szymanska E, Ciosek P, Winnicka K (2016) Alginate: current use and future perspectives in pharmaceutical and biomedical applications. Int J Poly, Sci
6. Andersen T, Melvik JE, Gåserød O, Alsberg E, Christensen BE (2012) Correction to ionically gelled alginate foams: physical properties controlled by operational and macromolecular parameters. Biomacromol 13:3703–3710
7. Fischer FG, Dorfel H (1955) Polyuronic acids in brown algae. Hoppe-Seyler's Z Physiol Chem 302:186–203
8. Lee KY, Mooney DJ (2012) Alginate: properties and biomedical applications. Prog Poly Sci 37:106–126
9. Beavis' A, Charlier RH (1987) An economic appraisal for the onshore cultivation of Laminaria spp. Hydrobiologia 151(152):387–398

10. Shroff P, Parikh S (2018) Production and characterization of alginate extracted from Paenibacillusriograndensis. Res J Life Sci Bioinfo Pharm Chem Sci 560–575
11. Saude N, Chèze-Lange H, Beunard D, Dhulster P, Guillochon D, Cazé A, Morcellet M, Junter G (2002) Alginate production by Azotobactervinelandii in a membrane bioreactor. Proc Biochem 38:273–278
12. McHugh DJ (2003) A guide to the seaweed industry. http://www.fao.org/3/a-y4765e.pdf
13. Yuan Y, Macquarrie D (2015) Microwave assisted extraction of sulfated polysaccharides (fucoidan) from Ascophyllumnodosum and its antioxidant activity. Carbohyd Polym 129:101–107
14. Hernández-Carmona G, McHugh DJ, López-Gutiérrez F (1999) Pilot plant scale extraction of alginates from Macrocystispyrifera 2 studies on extraction conditions and methods of separating the alkaline-insoluble residue. J App Phycol 11:493–502
15. Lüning K (1980) Critical levels of light and temperature regulating the gametogenesis of three Laminaria species (Phaeophyceae). J Phycol 16:1–15
16. Liu F, Shao Z, Zhang H, Liu J, Wang X, Duan D (2010) QTL Mapping for frond length and width in Laminaria japonica Aresch (Laminarales, Phaeophyta) using AFLP and SSR Markers. Mar Biotechnol 12:386–394
17. Taylor WR (1957) Marine algae of the northeastern coast of North America. University of Michigan Press, Ann Arbor MI, USA
18. Marins BV, Amado-Filho GM, Barreto MBB, Longo LL (2012) Taxonomy of the southwestern Atlantic endemic kelp: Laminariaabyssalis and Laminariabrasiliensis (Phaeophyceae, Laminariales) are not different species. Phycol Res 60:51–60
19. Westermeier R, Gomez I (1996) Biomass, energy contents and major organic compounds in the brown alga Lessonia nigrescens (Laminariales, Phaeophyceae) from Mehuin, South Chile. Bot Mar 39:553–559
20. Deysher LE, Dean TA (1986) *In situ* recruitment of sporophytes of the giant kelp, Macrocystispyrijkra (L.) C.A. Agardh: effects of physical factors. J Exp Mar Biol Ecol 103:41–63
21. Silva M, Gomes F, Oliveira F, Morais S, Delerue-Matos C (2015) Microwave-assisted alginate extraction from Portuguese Saccorhizapolyschides–influence of acid pretreatment. Int J Chem Mol Nucl Mat Metal Engg 9:1
22. Anderson RJ, Carrick P, Levitt GJ, Share A (1997) Holdfasts of adult kelp Ecklonia maxima provide refuges from grazing for recruitment of juvenile kelps. Mar Ecol Prog Ser 159:265–273
23. Ahn BR, Moon HE, Kim HR, Jung HA, Choi JS (2012) Neuroprotective effect of edible brown alga Eiseniabicyclis on amyloid beta peptide induced toxicity in PC12 cells. Arch Pharm Res 35:1989–1998
24. Chapman VJ (2009) A contribution to the ecology of Egregialaevigata Setchell I. taxonomic status and morphology. Botanica Marina 3(2)
25. Deckert RJ, Garbary DJ (2005) Ascophyllum and its symbionts.VIII. Interactions among Ascophyllumnodosum (Phaeophyceae), Mycophyciasascophylli (Ascomycetes) and Elachistafucicola (Phaeophyceae). Algae 20:363–368
26. Westermeier' R, Iiller DG, Gomezl I, Ivera P, Enzel H (1994) Population biology of Durvillaeaantarctica and Lessonianigrescens (Phaeophyta) on the rocky shores of southern Chile. Mar Ecol Prog Ser 110:187–194
27. Küpper FC, Peters AF, Shewring DM, Sayer MDJ, Mystikou A, Brown H, Azzopardi E, Dargent O, Strittmatter M, Brennan D, Asensi AO, van West P, Wilce RT (2016) Arctic marine phytobenthos of northern Baffin island. J Phycol 52:532–549
28. Rajauria G, Jaiswal AK, Abu-Gannam N, Gupta S (2013) Antimicrobial, antioxidant and free radical-scavenging capacity of brown seaweed Himanthaliaelongata from western coast of Ireland. J Food Biochem 37:322–335
29. McKenzie PF, Bellgrove A (2008) Dispersal of Hormosirabanksii (phaeophyceae) via detached fragments: reproductive viability and longevity. J Phycol 44:1108–1115

30. Parente MI, Neto AI (2003) Morphology and life history of Scytosiphonlomentaria (Scytosiphonaceae, Phaeophyceae) from the Azores. J Phycol 39:353–359
31. Dhargalkar VK, Deshmukhe GV (1996) Subtidal marine algae of the Dwaraka Cost (Gujrat). Indian J Marine Sci 25:297–301
32. Trujillo-Roldan MA, Moreno S, Espn G, Galindo E (2004) The roles of oxygen and alginate-lyase in determining the molecular weight of alginate produced by Azotobactervinelandii. Appl Microbiol Biotechnol 63:742–747
33. Ramsey DM, Wozniak DJ (2005) Understanding the control of Pseudomonas aeruginosa alginate synthesis and the prospects for management of chronic infections in cystic fibrosis. Mol Microbiol 56:309–322
34. Conti E, Flaibani A, O'Regan M, Sutherland' IW (1994) Alginate from Pseudornonasfluorescens and P putida: production and properties. Microbiol 140:1125–1132
35. Müller JM, Alegre RM (2007) Alginate production by Pseudomonas mendocina in a stirred draft fermenter. World J Microbiol Biotechnol 23:691–695
36. Fakhr MK, Va´Zquez AP, Chakrabarty AM, Bender CL (1999) Regulation of alginate biosynthesis in Pseudomonas syringaepv. Syringae. J Bacteriol 181:3478–3485
37. Torres MR, Sousa APA, Filho EAT, Melo DF, Feitosa JPA, Paulab RCMD, Lima MGS (2007) Extraction and physicochemical characterization of Sargassumvulgare alginate from Brazil. Carbohyd Res 342:2067–2074
38. Andriamanantoanina H, Rinaudob M (2010) Characterization of the alginates from five madagascan brown algae. Carbohyd Polym 82:555–560
39. Calumpong PH, Maypa PA, Magbanua M (1999) Population and alginate yield and quality assessment of four Sargassum species in Negros Island, central Philippines. Hydrobiologia 398:211–215
40. Fertah M, Belfkira A, Dahmane EM, Taourirte M, Brouillette F (2017) Extraction and characterization of sodium alginate from Moroccan Laminariadigitata brown seaweed. Arab J Chem 10:S3707–S3714
41. Urtuvia V, Maturana N, Acevedo F, Peña C, Barrera AD (2017) Bacterial alginate production: an overview of its biosynthesis and potential industrial production. World J Microbiol Biotechnol 33:198
42. Haug A, Larsen B, Smidsrød O (1966) A study of the constitution of alginic acid by partial hydrolysis. Acta Chem Scand 20:183–190
43. Penman A, Sanderson GR (1972) A method for the determination of uronic acid sequence in alginates. Carbohyd Res 25:273–282
44. Grasdalen H, Larsen B, Smidsrød O (1979) PMR study of the composition and sequence of uronate residues in alginates. Carbohyd Res 68:23–31
45. Smidsrød O, Haug A, Larsen B (1966) The influence of pH on the rate of hydrolysis ofacidic polysaccharides. Acta Chem Scand 20:1026–1034
46. Tsujino I, Saito T (1961) A new unsaturated uronide isolated from alginasehydrolysate. Nature 192:970–971
47. Preiss J, Ashwell G (1962) Alginic acid metabolism in bacteria. J Biol Chem 237:309–316
48. Haug AL, Larsen B, Smidsrød O (1963) The degradation of alginates at different pH values. Acta Chem Scand 17:1466–1468
49. Haug AL, Larsen B, Smidsrød O (1967) Alkaline degradation of alginate. Acta Chem Scand 21:2859–2870
50. Smidsrød O, Haug A, Larsen B (1963) The influence of reducing substances on the rate of degradation of alginates. Acta Chem Scand 17:1473–1474
51. Smidsrød O, Haug A, Larsen B (1963) Degradation of alginate in the presence of reducing compounds. Acta Chem Scand 17:2628–2637
52. Leo WJ, Mcloughlin AJ, Malone DM (1990) Effects of sterilization treatments on some properties of alginate solutions and gels. Biotechnol Prog 6:51–53
53. Grasdalen H (1983) High-field, ^{1}H-NMR spectroscopy of alginate: sequential structure and linkage conformations. Carbohyd Res 118:255–260

54. Zheng N, Zhang Y, Fan X, Han L (1994) Effects of composition and structure of alginates on adsorption of divalent metals. Chin J Oceanol Limnol 12:78–83
55. Rioux LE, Turgeon SL, Beaulieu M (2007) Characterization of polysaccharides extracted from brown seaweeds. Carbohyd Polym 69:530–537
56. Smidsrød O, Draget KI (1996) Alginates: Chemistry and Physical Properties. Carbohyd Eur 14:6–13
57. Skjik-Braek G (1992) Alginates: biosyntheses and some structure-function relationships relevant to biomedical and biotechnological applications. Biochem Plant Polysacch, 27–33
58. Hang A, Larsen B, Smidsrød O (1974) Uronic acid sequence in Alginate from different sources. Carbohyd Res 32:217–225
59. McAvoy MJ, Newton V, Paull A, Morgan J, Gacesa P, Russell NJ (1989) Isolation of mucoid strains of Pseudomonas aeruginosa from non-cystic-fibrosis patients and characterisation of the structure of their secreted alginate. J Med Microbiol 28:183–189
60. Tielen P, Strathmann M, Jaeger KE, Flemming HC, Wingender J (2005) Alginate acetylation influences initial surface colonization by mucoid Pseudomonas aeruginosa. Microbiol Res 160:165–176
61. Karri VVSR, Kuppusamy G, Talluri SV, Mannemala SS, Kollipara R, Wadhwani AD (2016) Curcumin loaded chitosan nanoparticles impregnated into collagen-alginate scaffolds for diabetic wound healing. Int J Biol Macromol 93:1519–1529
62. Gomez CG, Perez-Lambrecht MV, Lozano JE, Rinaudo M, Villar MA (2009) Influence of the extraction-purification conditions on final properties of alginates obtained from brown algae Macrocystispyrifera. Int J Biol Macromol 44:365–371
63. Bugaj J, Gorecki M (1995) Kinetics of dynamic viscosity changes of aqueous sodium carboxymethylcellulose and sodium alginate solutions. Pharmazie 50:750–752
64. Duggirala S, DeLuca PP (1996) Rheological characterization of cellulosic and alginate polymers. PDA J Pharm Sci Technol 50:290–296
65. Bugaj J, Gorecki M (1996) Rheometrical estimation of physical properties of sodium alginate and sodium carboxymethylcellulose aqueous solutions. Acta Pol Pharm Drug Res 53:141–146
66. Nayak AK, Pal D (2011) Development of pH-sensitive tamarind seed polysaccharide-alginate composite beads for controlled diclofenac sodium delivery using response surface methodology. Int J Biol Macromol 49:784–793
67. Yang JS, Xie YJ, He W (2011) Research progress on chemical modification of alginate: a review. Carbohyd Polym 84:33–39
68. Bodmeier R, Paeratakul O (1991) A novel multiple-unit sustained release indomethacin-hydroxypropyl methylcellulose delivery system prepared by ionotropic gelation of sodium alginate atelevated temperatures. Carbohyd Polym 16:399–408
69. Goh CH, Heng PWS, Chan LW (2012) Alginates as a useful natural polymer for microencapsulation and therapeutic applications. Carbohyd Polym 88:1–12
70. Shilpa A, Agarwal SS, Rao AR (2003) Controlled delivery of drugs from alginate matrix. J Macromol Sci Polym Rev 43:187–221
71. Nayak AK, Das B, Maji R (2012) Calcium alginate/gum Arabic beads containing glibenclamide: development and in vitro characterization. Int J Biol Macromol 51:1070–1078
72. George M, Abraham TE (2006) Polyionic hydrocolloids for the intestinal delivery of protein drugs: alginate and chitosan—a review. J Control Release 114:1–14
73. Reis CP, Neufeld RJ, Vilela S, Ribiero AJ, Veiga F (2006) Review and current status of emulsion/dispersion technology using an internal gelation process for the design of alginate particles. J Microencapsul 23:245–257
74. Wandrey C, Espinosa D, Rehor A, Hunkeler D (2003) Influence of alginate characteristics on the properties of multi-component microcapsules. J Microencapsul 20:597–611
75. Draget KI, Skja˚k-Braek G, Smidsrød O (1997) Alginate based new materials. Int J Biol Macromol 21:47–55
76. Bhowmik BB, Sa B, Mukherjee A (2006) Preparation and in vitrocharacterization of slow release testosterone nanocapsules in alginates. Acta Pharm 56:417–429

77. Patel YL, Sher P, Pawar AP (2006) The effect of drug concentration and curing time on processing and properties of calcium alginate beads containing metronidazole by response surfacemethodology. AAPS PharmSciTech 7:E24–E30
78. Al-Kassas R, Al-Gohary OMN, Al-Fadhel MM (2007) Controlling of systemic absorption of gliclazide through incorporation into alginate beads. Int J Pharm 341:230–237
79. Smrdel P, Bogataj M, Mehar A (2008) The influence of selected parameters on the size and shape of alginate beads prepared by ionotropic gelation. Sci Pharm 76:77–90
80. Jana S, Gandhi A, Sheet S, Sen KK (2015) Metal ion-induced alginate–locust bean gum IPN microspheres for sustained oral delivery of aceclofenac. Int J Biol Macromol 72:47–53
81. Racovita S, Vasilu S, Popa M, Luca C (2009) Polysaccharides based micro- and nanoparticles obtained by ionic gelation and their applications as drug delivery systems. Rev Rom De Chim 54:709–718
82. Sinha P, Ubaidulla U, Hasnain MS, Nayak AK, Rama B (2015) Alginate-okra gum blend beads of diclofenac sodium from aqueous template using $ZnSO_4$ as a cross-linker. Int J Biol Macromol 79:555–563
83. Sinha P, Ubaidulla U, Nayak AK (2015) Okra (*Hibiscus esculentus*) gum-alginate blend mucoadhesive beads for controlled glibenclamide release. Int J Biol Macromol 72:1069–1075
84. Nayak AK, Kalia S, Hasnain MS (2013) Optimization of aceclofenac-loaded pectinate-poly(vinylpyrrolidone) beads by response surface methodology. Int J Biol Macromol 62:194–202
85. Nayak AK, Pal D, Pradhan J, Ghorai T (2012) The potential of *Trigonella foenum-graecum* L. seed mucilage as suspending agent. Indian J Pharm Edu Res 46:312–317
86. Ghosh D, Pramanik A, Sikdar N, Pramanik P (2011) Synthesis of low molecular weight alginic acid nanoparticles through persulfate treatment as effective drug delivery system to manage drug resistant bacteria. Biotechnol Bioproc E 16:383–392
87. Sriamornsak P, Sungthongjeen S (2007) Modification of theophylline release with alginate gel formed in hard capsules. AAPS PharmSciTech 8:E1–E8
88. Grassi M, Colombo I, Lapasin R (2001) Experimental determination of the theophylline diffusion coefficient in swollen sodium-alginate membranes. J Control Release 76:93–105
89. Miyazaki S, Kubo W, Attwood D (2000) Oral sustained delivery of theophylline using *in-situ* gelation of sodium alginate. J Control Release 67:275–280
90. El-Ghaffar MAA, Hashem MS, El-Awady MK, Rabie AM (2012) pH-sensitive sodium alginate hydrogels for riboflavin controlled release. Carbohyd Polym 89:667–675
91. Se´choy O, Tissie´ G, Se´bastian C, Maurin F, Driot J, Trinquand C (2000) A new long acting ophthalmic formulation of carteolol containing alginic acid. Int J Pharm 207:109–116
92. Makwana SB, Patel VA, Parmar SJ (2016) Development and characterization of *in-situ* gel for ophthalmic formulation containing ciprofloxacin hydrochloride. Res Pharm Sci 6:1–6
93. Mandal S, Thimmasetty KMJM, Prabhushankar GL, Geetha MS (2012) Formulation and evaluation of an *in situ* gel-forming ophthalmic formulation of moxifloxacin hydrochloride. Int J Pharm Investig 2:78–82
94. Draget KI, Skjik-Braek G, Christensen BE, Gåserød O, Smidsrød O (1996) Swelling and partial solubilization of alginic acid gel beads in acidic buffer. Carbohyd Polym 29:209–215
95. Ali A, Ahmed S (2018) A review on chitosan and its nanocomposites in drug delivery. Int J Biol Macromol 109:273–286
96. Nagahama H, Nwe N, Jayakumar R, Koiwa S, Furuike T, Tamura H (2008) Novel biodegradable chitin membranes for tissue engineering applications. Carbohyd Polym 73:295–302
97. Wang Y, Chen H, Wang J, Xing L (2014) Preparation of active corn peptides from zein through double enzymes immobilized with calcium alginate–chitosan beads. Process Biochem 49:1682–1690
98. Takka S, Gürel A (2010) Evaluation of chitosan/alginate beads using experimental design: Formulation and *in vitro* characterization. AAPS PharmSciTech 11:460–466
99. Zhang Y, Wei W, Lv P, Wang L, Ma G (2011) Preparation and evaluation of alginate–chitosan microspheres for oral delivery of insulin. Eur J Pharm Biopharm 77:11–19

100. Lacerda L, Parize AL, Fávere V, Laranjeira MCM, Stulzer HK (2014) Development and evaluation of pH-sensitive sodium alginate/chitosan microparticles containing the antituberculosis drug rifampicin. Mater Sci Eng, C 39:161–167
101. Coppi G, Iannuccelli V, Sala N, Bondi M (2004) Alginate microparticles for Polymyxin B Peyer's patches uptake: microparticles for antibiotic oral administration. J Microencapsul 21:829–839
102. Čalija B, Milić J, Cekić N, Krajišnik D, Daniels R, Savić S (2013) Chitosan oligosaccharide as prospective cross-linking agent for naproxen-loaded Ca-alginate microparticles with improved pH sensitivity. Drug Dev Ind Pharm 39:77–88
103. Onishi H, Koyama K, Sakata O, Machida Y (2010) Preparation of chitosan/alginate/calcium complex microparticles loaded with lactoferrin and their efficacy on carrageenan-induced edema in rats. Drug Dev Ind Pharm 36:879–884
104. Li P, Dai Y-N, Zhang J-P, Wang A-Q, Wei Q (2008) Chitosan-alginate nanoparticles as a novel drug delivery system for nifedipine. Int J Biomed Sci 4:221–228
105. Patil JS, Kamalapur MV, Marapur SC, Shiralshetti SS, Kadam DV (2012) Ionotropically gelled chitosan-alginate complex hydrogel beads: Preparation, characterization and *in-vitro* evaluation. Ind J Pharm Edu Res 46:248–252
106. Dionísio M, Grenha A (2012) Locust bean gum: Exploring its potential for biopharmaceutical applications. J Pharm Bioallied Sci 4:175–185
107. Prajapati VD, Jani GK, Moradiya NG, Randeria NP, Nagar BJ (2013) Locust bean gum: a versatile biopolymer. Carbohyd Polym 94:814–821
108. Alves M, Antonov YA, Goncalves M (1999) The effect of structural features of gelatin on its thermodynamic compatibility with locust bean gum in aqueous media. Food Hydrocoll 13:157–166
109. Upadhyay M, Reddy SKA, Vardhan H, Yadav SK, Mishra B (2019) Locust bean gum and sodium alginate based interpenetrating polymeric network microbeads encapsulating capecitabine: Improved pharmacokinetics, cytotoxicity & in vivo antitumor activity. Mater Sci Eng, C 104:
110. Pawar HA, Lalitha KG, Ruckmani K (2015) Alginate beads of captopril using galactomannan containing sennatora gum, guar gum and locust bean gum. Int J Biol Macromol 76:119–131
111. Dey P, Sa B, Maiti S (2015) Impact of gelation period on modified locust bean-alginate interpenetrating beads for oral glipizide delivery. Int J Biol Macromol 76:176–180
112. Glicksman M (1986) Tamarind seed gum. In: Glicksman M (ed) Food Hydrocolloids, vol 3. CRC Press, Boca Raton, FL, pp 191–202
113. Yamatoya K, Shirakawa M (2003) Xyloglucan: structure, rheological properties, biological functions and enzymatic modification. Curr Trend Polym Sci 8:27–72
114. Doniparthi J, Jyothi BJ (2020) Novel tamarind seed gum-alginate based multi-particulates for sustained release of dalfampridine using response surface methodology. Int J Biol Macromol 144:725–741
115. Pal D, Nayak AK (2012) Novel tamarind seed polysaccharide-alginate mucoadhesive microspheres for oral gliclazide delivery: *in vitro–in vivo* evaluation. Drug Deliv 19:123–131
116. Jana S, Sharma R, Maiti S, Sen KK (2016) Interpenetrating hydrogels of O-carboxymethyl Tamarind gum andalginate for monitoring delivery of acyclovir. Int J Biol Macromol 92:1034–1039
117. Bera H, Boddupalli S, Nandikonda S, Kumar S, Nayak AK (2015) Alginate gel-coated oil-entrapped alginate–tamarind gum–magnesium stearate buoyant beads of risperidone. Int J Biol Macromol 78:102–111
118. Nayak AK, Pal D, Santra K (2016) Swelling and drug release behavior of metformin HCl-loaded tamarind seed polysaccharide-alginate beads. Int J Biol Macromol 82:1023–1027
119. Doi Y, Ichihara T, Hagiwara A, Imai N, Tamano S, Orikoshi H, Ogasawara K, Sasaki Y, Nakamura M, Shirai T (2006) A ninety-day oral toxicity study of a new type of processed gum arabic, from Acacia tree (*Acacia senegal*) exudates, in F344 rats. Food Chem Toxicol 44:560–566

120. Verbeken D, Dierckx S, Dewettinck K (2003) Exudate gums: occurrence, production, and applications. Appl Microbiol Biotechnol 63:10–21
121. Gils PS, Ray D, Sahoo PK (2010) Designing of silver nanoparticles in gum arabic based semi-IPN hydrogel. Int J Biol Macromol 46:237–244
122. Li Y, Liang M, Dou X, Feng C, Pang J, Cheng X, Liu H, Liu T, Wang Y, Chen X (2019) Development of alginate hydrogel/gum Arabic/gelatin based composite capsules and their application as oral delivery carriers for antioxidant. Int J Biol Macromol 132:1090–1097
123. Tsai FH, Kitamura Y, Kokawa M (2017) Effect of gum arabic-modified alginate on physicochemical properties, release kinetics, and storage stability of liquid-core hydrogel beads. Carbohyd Polym 174:1069–1077
124. Mohamed HN, Mustafa S, Fitrianto A, Manap YA (2017) Development of alginate–gum arabic beads for targeted delivery of protein. J Biomol Res Ther 6:1000155
125. Kono H, Erata T, Takai M (2002) CP/MAS ^{13}C NMR study of cellulose and cellulose derivatives. 2. complete assignment of the ^{13}C resonance for the ring carbons of cellulose triacetate polymorphs. J Am Chem Soc 124:7512–7518
126. Vuoti S, Laatikainen E, Heikkinen H, Johansson L-S, Saharinen E, Retulainen E (2013) Chemical modification of cellulosic fibers for better convertibility in packaging applications. Carbohyd Polym 96:549–559
127. Sun B, Zhang M, Shen J, He Z, Fatehi P, Ni Y (2019) Applications of cellulose-based materials in sustained drug delivery systems. Curr Med Chem 26:2485–2501
128. Kim MS, Park SJ, Gu BK, Kim C-H (2012) Ionically crosslinked alginate–carboxymethyl cellulose beads for the delivery of protein therapeutics. Appl Surf Sci 262:28–33
129. Surrao DC, Arasu Y, Ekberg JAK, John JAS (2020) Blended, crosslinked alginate-methylcellulose hydrogels for encapsulation and delivery of olfactory ensheathing cells. Materialia 10:
130. Sharma S, Sarkar G, Srestha B, Chattopadhyay D, Bhowmik M (2019) *In-situ* fast gelling formulation for oral sustained drug delivery of paracetamol to dysphagic patients. Int J Biol Macromol 134:864–868
131. Angadi SC, Manjeshwar LS, Aminabhavi TM (2012) Novel composite blend microbeads of sodium alginate coated with chitosan for controlled release of amoxicillin. Int J Biol Macromol 51:45–55
132. Bera H, Kandukuri SG, Nayak AK, Boddupalli S (2015) Alginate-sterculia gum gel-coated oil-entrapped alginate beads for gastroretentive risperidone delivery. Carbohyd Polym 120:74–84
133. Singh B, Pal L (2008) Development of sterculia gum based wound dressings for use in drug delivery. Eur Polym J 44:3222–3230
134. Nath B, Nath LK (2013) Design, development, and optimization of sterculia gum-based tablet coated with chitosan/eudragit RLPO mixed blend polymers for possible colonic drug delivery. J Pharm 546324 (2013). https://doi.org/10.1155/2013/546324
135. Chivate AA, Poddar SS, Abdul S, Savant G (2008) Evaluation of Sterculia foetida gum as controlled release excipient. AAPS PharmSciTech 9:197–204
136. Guru PR, Nayak AK, Sahu RK (2013) Oil-entrapped sterculia gum–alginate buoyant systems of aceclofenac: Development and in vitro evaluation. Colloid Surface B. 104:268–275
137. Singh B, Sharma V, Chauhan D (2010) Gastroretentive floating sterculia–alginate beads for use in antiulcer drug delivery. Chem Eng Res Des 88:997–1012
138. Md S, Ahuja A, Khar RK, Baboota S, Chuttani K, Mishra AK, Ali J (2011) Gastroretentive drug delivery system of acyclovir-loaded alginate mucoadhesive microspheres: Formulation and evaluation. Drug Deliv 18:255–264
139. Qurrat-ul-Ain, Sharma S, Khuller GK, Garg SK (2003) Alginate-based oral drug delivery system for tuberculosis: pharmacokinetics and therapeutic effects. J Antimicrob Chemother 51:931–938
140. Malakar J, Nayak AK, Pal D (2012) Development of cloxacillin loaded multiple-unit alginate-based floating system by emulsion–gelation method. Int J Biol Macromol 50:138–147
141. Bonilla P, Arias EM, Solans C, García-Celma MJ (2018) Influence of crosslinked alginate on drug release from highly concentrated emulsions. Colloid Surface A 536:148–155

142. Ramana BV, Sana SJ, Swapna LC, Sekhar SC, Ademma G, Murthy TEGK (2016) Design and development of floating *in-situ* gel of pantoprazole. Der Pharma Lett 8:239–249
143. Wang C, Liu H, Gao Q, Liu X, Tong Z (2008) Alginate–calcium carbonate porous microparticle hybrid hydrogels with versatile drug loading capabilities and variable mechanical strengths. Carbohyd Polym 71:476–480
144. Jivani RR, Patel CN, Patel DM, Jivani NP (2010) Development of a novel floating in-situ gelling system for stomach specific drug delivery of the narrow absorption window drug baclofen. Iran J Pharm Res 9:359–368
145. Soumia A, Adel M, Amina S, Bouhadjar B, Amal D, Farouk Z, Abdelkader B, Mohamed S (2020) Fe_3O_4-alginate nanocomposite hydrogel beads material: One-pot preparation, release kinetics and antibacterial activity. Int J Biol Macromol 145:466–475
146. Amiri M, Salavati-Niasari M, Pardakhty A, Ahmadi M, Akbari A (2017) Caffeine: A novel green precursor for synthesis of magnetic $CoFe_2O_4$ nanoparticles and pH-sensitive magnetic alginate beads for drug delivery. Mater Sci Eng C 76:1085–1093
147. Liu H, Wang C, Gao Q, Chen J, Ren B, Liu X, Tong Z (2009) Facile fabrication of well-defined hydrogel beads with magnetic nanocomposite shells. Int J Pharm 376:92–98
148. Supramaniam J, Adnan R, Kaus NHM, Bushra R (2018) Magnetic nanocellulose alginate hydrogel beads as potential drug delivery system. Int J Biol Macromol 118:640–648
149. Rashidzadeh B, Shokri E, Mahdavinia GR, Moradi R, Mohamadi-Aghdam S, Abdi S (2020) Preparation and characterization of antibacterial magnetic-/pH sensitive alginate/Ag/Fe_3O_4 hydrogel beads for controlled drug release. Int J Biol Macromol 154:134–141
150. Lei H, Xie M, Zhao Y, Zhang F, Xu Y, Xie J (2016) Chitosan/sodium alginate modificated grapheme oxide-based nanocomposite as a carrier for drug delivery. Ceram Int 42:17798–17805
151. Veerla SC, Kim DR, Yang SY (2018) Retraction note: Fabrication of a microfluidic device for studying the *in situ* drug-loading/release behavior of graphene oxide-encapsulated hydrogel beads. Biomater Res 22:15. https://doi.org/10.1186/s40824-018-0125-y

Chapter 4
Ionic Gelled Chitosan for Drug Delivery

Supriyo Saha and Dilipkumar Pal

Abstract The natural polysaccharide obtained from crustacean orthopods like shrimp, lobster, crabs followed by demineralization, decolorization, and deacetylation, known as chitosan which was reacted with tripolyphosphate as crosslinking agent with pH adjustment obtain micro/nanoparticle. This technique was a way to deliver a formulation with improved surface topology, particle size, polydispersity index, and modified surface charge for attachment of functional groups of positively charged polymer as chitosan, that have been used for drug delivery of MDA-MB-231, doxorubicin to treat breast/skin/colorectal cancer, bovine serum albumin, *Saturejahortensis* essential oil to treat acaricide, of zinc in MACS 3125 and UC1114 wheat cultivars, particular gene, tacrine, amoxicillin to treat *Helicobacter pylori*, luciferase PVAX1-Luc with in vitro transfection efficiency and also these nanoparticles showed inhibition of *Aedes aegypti*, *Candida albicans* with wound healing and mercury adsorption properties. The chapter describes preparations, physicochemical properties, and drug delivery applications of various ionically gelled chitosan.

Keywords Chitosan · Sodium tripolyphosphate · Drug delivery · Acaricidal · Antifungal · Anticancer · Peptic ulcer · Gene delivery

1 Introduction

Chitosan was discovered from the shells of crustacean orthopods like shrimp, lobster, crabs, etc. [15, 29]. Then the shells were treated with lactic acid or *Pseudomonas aeruginosa* followed by treatment with protease enzyme or base treatment to make it free from mineral and protein, respectively [19, 27, 28]. Then the extract was reacted with acetone to became colorless followed by reacting with deacetylase enzyme to

S. Saha
School of Pharmaceutical Sciences and Technology, Sardar Bhagwan Singh University, Dehradun 248161, Uttarakhand, India

D. Pal (✉)
Department of Pharmaceutical Sciences, Guru Ghasidas Vishwavidyalaya (A Central University), Bilaspur 495009, Chhattisgarh, India

A. K. Nayak et al. (eds.), *Ionically Gelled Biopolysaccharide Based Systems in Drug Delivery*, Gels Horizons: From Science to Smart Materials,
https://doi.org/10.1007/978-981-16-2271-7_4

form a long-chain polymer of N-acetylglucosamine with more than 30% degree of deacetylation (Fig. 1) [18, 20, 23].

Chitosan was linked via β-1,4-glucosamine linkges (Fig. 2).

Sodium tripolyphosphate ($Na_5P_3O_{10}$)was an inorganic stoichiometric mixture of disodium hydrogen phosphate, monosodium dihydrogenorthophosphate, existed as colorless hexahydrate form (Fig. 3) [26–28].

Sodium tripolyphosphate was created strong bond with metal using two or three co-ordinate linkages. The tripolyphospate group was caused inhibition of growth of *Streptococcus mutans* with greater complexomteric and bacterial inhibitory activity [17] (Table 1).

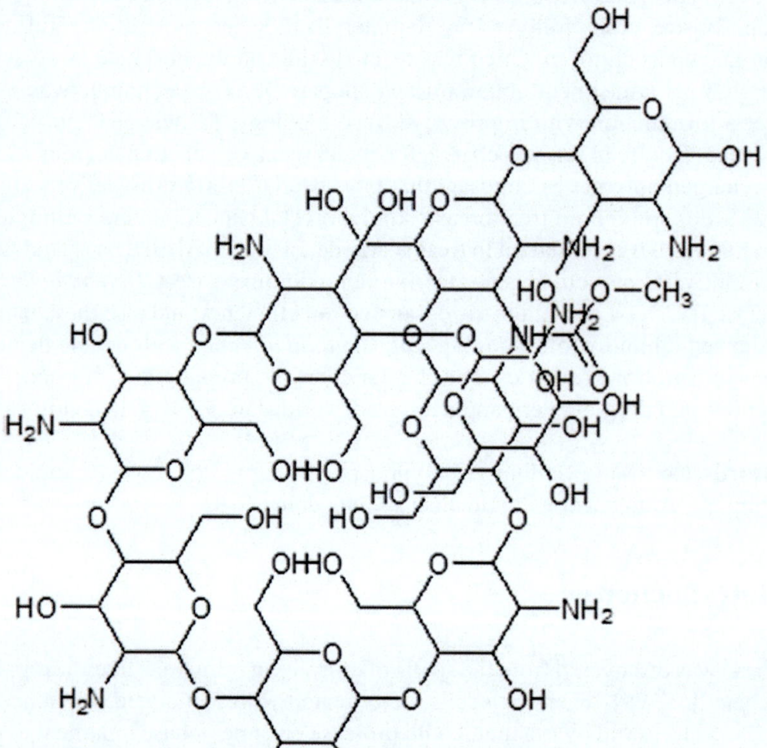

Fig. 1 Structure of chitosan with PubChem ID: 71853

Fig. 2 Production of chitosan

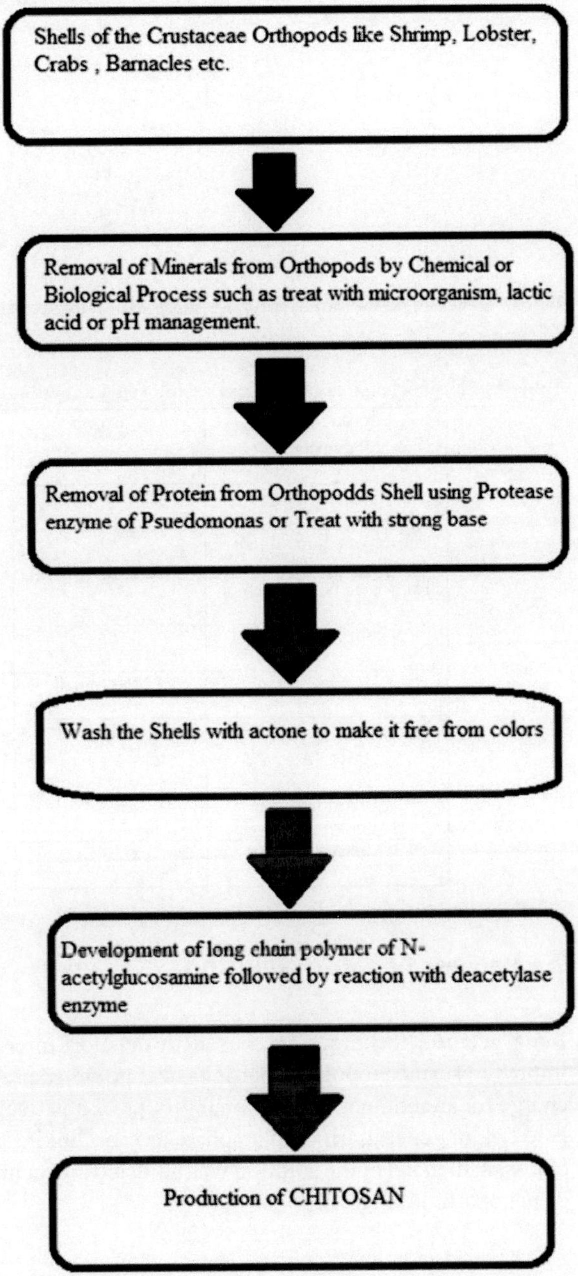

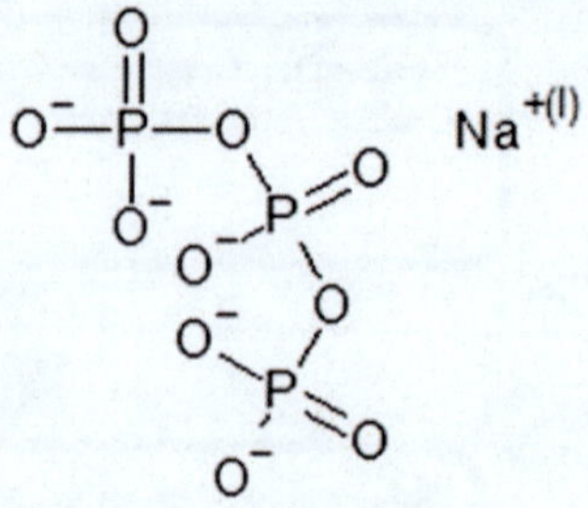

Fig. 3 Structure of sodium tripolyphosphate with Pubchem ID: 24455

Table 1 Physicochemical Properties of Chitosan and Sodium Tripolyphosphate

Physicochemical Parameters	Chitosan	Sodium Tripolyphosphate
Molecular weight	1526.5 g/mol	376.86 g/mol
Hydrogen bond donor	29	0
Hydrogen bond acceptor	47	10
Rotatable Bond Count	27	2
Mass	1526.6 g/mol	367.81 g/mol
Topological Surface Area (Å²)	808	185
Heavy atom count	104	18
Melting point	220 °C	622°C

2 Process of Ionic Gelation

Ionic gelation technique was used to develop microparticle or nanoparticle with improved surface topology, particle size, polydispersity index, and modified surface charge for attachment of functional groups of positively charged polymer as chitosan [11, 12]. In general, tripolyphosphate as crosslinking agent was added to solution of chitosan, then pH of the solution was adjusted and centrifuged to obtain microparticle or nanoparticle (Fig. 4).

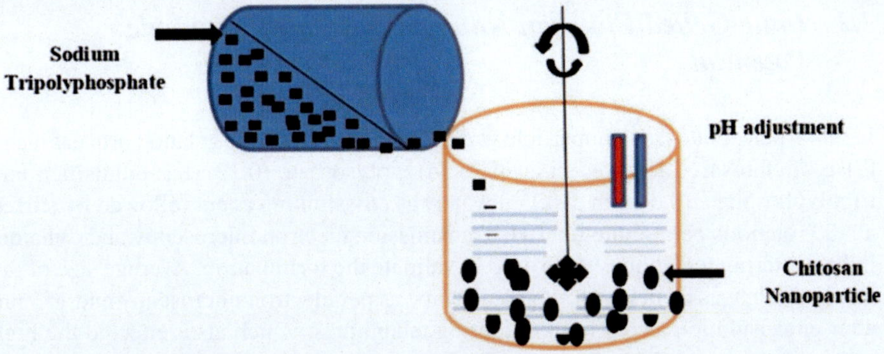

Fig. 4 Graphical representation of chitosan-tripolyphosphate ionic gelation nanoparticle

3 Application of Ionic Gelled Chitosan in Drug Delivery

3.1 Ionic Gelled Chitosan Nanoparticles for Drug Delivery

Here, chitosan nanoparticles were developed using ionic gelation process with tripolyphosphate as cross linking agent. The average size and zeta potential of the nanoparticles were 150–250 nm and 30 milivolt. The nanoparticle with 250 μg/ml concentration was observed with greater cell viable activity towards human breast cancer cell line (MDA-MB-231), which directly correlated with the efficient drug delivery [13].

3.2 Ionic Gelled Chitosan Nanoparticles with Efficient Protein Delivery

In this nanoparticle, various concentrations of chitosan (0.054, 0.069, 0.085, and 0.1%) with different pH values (3, 4, and 5) and similar concentrations of tripolyphosphate concentrations were crosslinked at 250 rotations per minute at 25 °C temperature for 30 min. The formulations were analyzed using transmission electron microscopy and average particle.

Size diameter was 250–350 nm with no proper change in zeta potential was observed with variable pH values. Here, bovine serum albumin was considered as model drug with 96.3% drug loading efficiency with chitosan and tripolyphosphate at 0.1% concentration at pH 5.0; also the protein-loaded nanoparticle was observed with 480 nm particle diameter (Fig. 5). This value indicates the importance of pH and chitosan/crosslinker concentration to develop an ideal nanoparticle [1].

3.3 Ionic Gelled Chitosan Nanoparticles with Acaricide Potential

In this article, chitosan nanoparticle was developed using ionic gelation process using 0.1 g of chitosan (dissolved in acetic acid), polysorbate (0.12 g) as emulsifier, and tripolyphosphate (0.04 g in distilled water) as crosslinking agent followed by stirred at 500 rotations per minute for 1 H. Transmission electron microscopy and dynamic light scattering techniques were used to validate the formulation. Average size of the nanoparticle was 80 nm with spherical shape as per electron microscopy and 153 nm after encapsulation as per light scattering techniques, which also reflected the high encapsulation (96.17%) efficiency. The acaricide activity with the observation the lethal concentration$_{50\%}$ of 4.95 μL/L& 46.98 μL/L and 2.02 μL/L & 31.30μL/L with *Satureja hortensis* essential oil (pure) and oil loaded nanoparticle, respectively after 1 days and 3 days. Also, after 18th day of pure oil and oil-loaded nanoparticles observed with 2.0% and 67.0% mortality rate, respectively (Fig. 6).

This data indicates greater inhibitions of *Tetranychus urticae* after 3 days of proper exposure with lethal concentration$_{50\%}$ of 107.38 μL/L [2].

3.4 Ionic Gelled Chitosan Nanoparticles with Larvacidal Potential

Here, nanoparticle was developed by reaction between 0.01% w/v chitosan solution (in acetic acid) and 0.1% w/v of sodium tripolyphosphate followed by magnetically stirred at 1000 rotations per minute for 15 min. The formulation was characterized by scanning/transmission electron microscopy, X-ray diffraction techniques. The

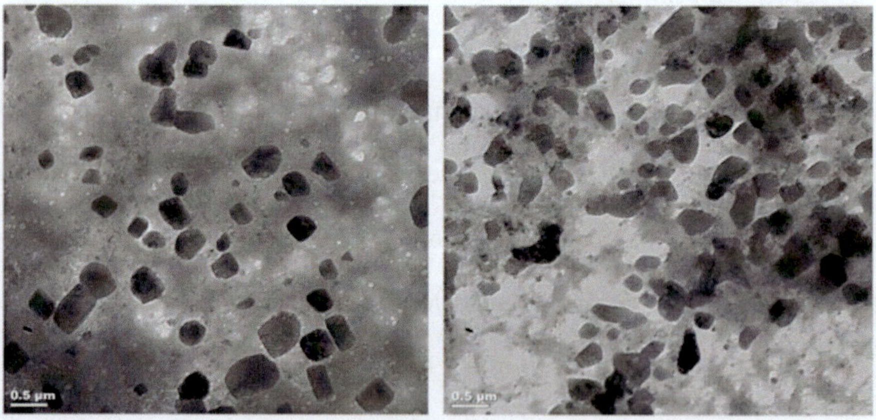

Fig. 5 TEM micrographs of (i) BSA-loaded chitosan/TPP; (ii) BSA-loaded chitosan/HMP nanoparticles. *Source* Abdelrahman and Hudson [1] with permission from Elsevier B.V.

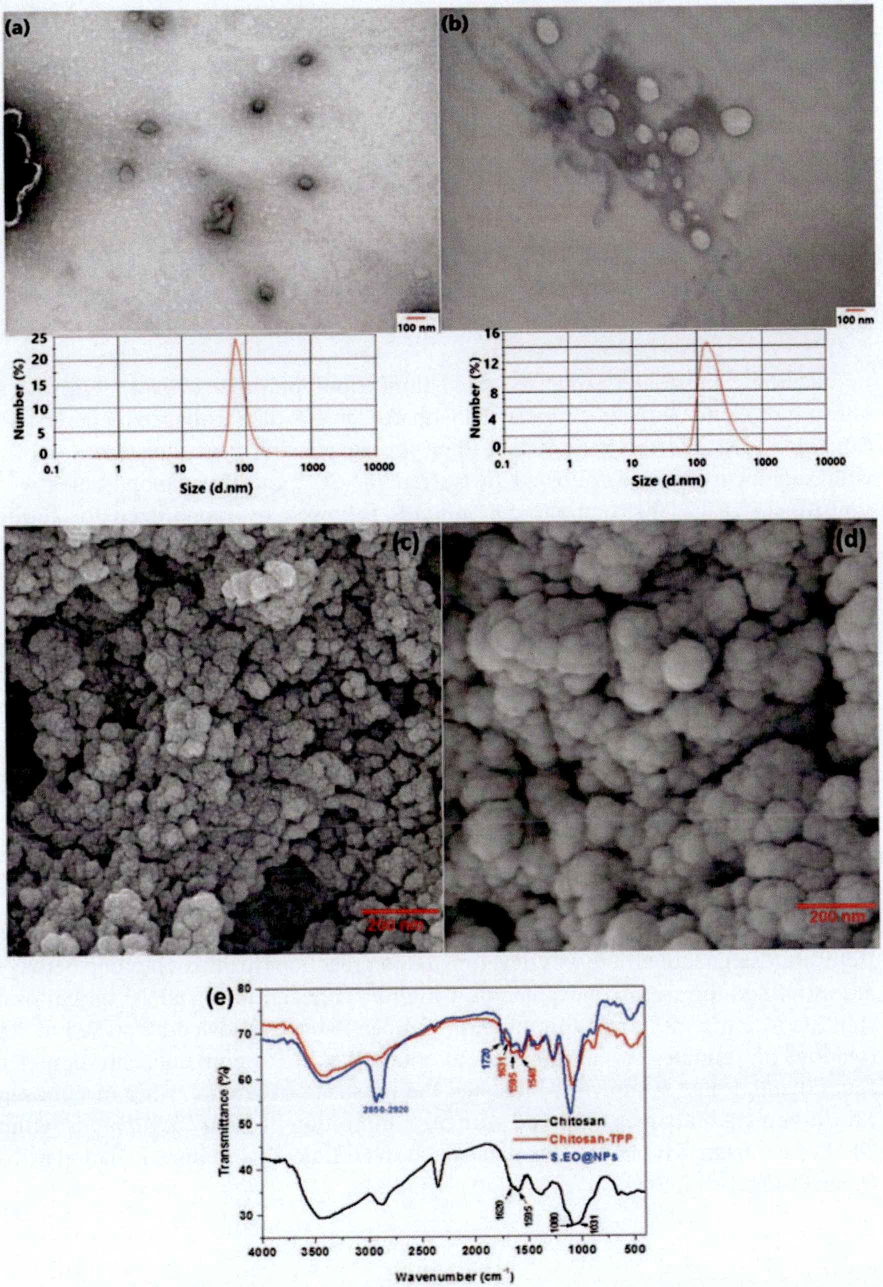

Fig. 6 **a** TEM and DLS results of (a) pure chitosan-TPP and **b** S.EO@NPs nanoparticle loaded with Satureja hortensis essential oil; SEM micrographs of **c** purechitosan-TPP and **d** S.EO@NPs nanoparticle; **e** FTIR spectra of pure chitosan, chitosan-TPP nanoparticles, and S.EO@NPs. *Source* Ahmadi et al. [2] with permission from Elsevier Inc.

formulations were observed with 8 nm (lower size) and 80 nm (larger particle size) as per electron microscopy and concentration (92.58 mg/L) observed with greater inhibition *Aedes aegypti* yellow fever causing mosquito larvae. This data clearly stated the importance of ionic gelled chitosan nanoparticles with larvacidal property [3].

3.5 Ionic Gelled Chitosan Nanoparticles with Anticancer Drug Delivery

In this article, chitosan (2 mg/ml) and sodium tripolyphosphate (0.84 mg/ml) were ultrasonicated to prepare chitosan nanaoparticle and then collagen peptide (0–2 %) was added into the nanoparticle then the suspension was adjusted to pH 7.0 with sodium hydroxide followed by stirred for 30 min. The nanoparticles were centrifuged at 12,000 rotations per minutes followed by lyophilized for further addition of doxorubicin (1 mg/ml) concentration in nanoparticle. The zeta potential of −26 milivolt indicated the stability of the formulation and size of 260 nm reflected the nanosize. The release of doxorubicin was followed by the two phasic system as burst release within 20 H and sustained release for 7 days. The pH 7.4 was observed with 68% release of the drug followed by 89% release at pH 1.5. This data indicated the efficient delivery of anticancer agent by collagen peptide linked chitosan nanoparticle [4].

3.6 Ionic Gelled Chitosan Nanoparticles with Antifungal Efficiency

Here, chitosan nanoparticle was developed upon reaction by infusion pump between 3.0 ml of sodium tripolyphosphate (2.4 mg/ml) concentration and 10 ml chitosan solution (4.0 mg/ml) concentration at 60 ml per H rate, followed by stirred at 700 rotations per minutes for 30 min span to obtain the 3.63 mg/ml concentration. The zeta potential of +50 milivolt indicated the presence of amino group of chitosan. The biological outcomes observed with 50% inhibition of *Candida albicans* within 24 H of exposure (Table 2). This data stated the efficiency of chitosan nanoparticle against fungal strain [5].

Table 2 Inhibitory effects of TPP/Chit nanoparticles, chitosan and nystatin on C. albicans initial adhesion and biofilm formation (24 h) after 1 min exposure time. The values are expressed as percentage of inhibition (I%)

Concentration	Initial adhesion inhibition			Inhibition of biofilm formation		
	TPP/Chit	Nystatin	Chitosan	TPP/Chit	Nystatin	Chitosan
MIC	$25 < I\% \leq 50^{Aa}$	$25 < I\% \leq 50^{Aa}$	$25 < I\% \leq 50^{Aa}$	$25 < I\% \leq 50^{Ba}$	$25 < I\% \leq 50^{Aa}$	$25 < I\% \leq 50^{Aa}$
2 MIC	$25 < I\% \leq 50^{Aa}$	$25 < I\% \leq 50^{Aa}$	$25 < I\% \leq 50^{Aa}$	$25 < I\% \leq 50^{Ba}$	$25 < I\% \leq 50^{Aa}$	$25 < I\% \leq 50^{Aa}$
4 MIC	$25 < I\% \leq 50^{Aa}$	$25 < I\% \leq 50^{Aa}$	$25 < I\% \leq 50^{Aa}$	$25 < I\% \leq 50^{Aa}$	$25 < I\% \leq 50^{Ab}$	$25 < I\% \leq 50^{Ab}$

*Different capital letters in each column indicate statistically significant differences (Kruskal-Wallis test followed by Dunn's post-hoc test, $p < 0.05$) among different concentrations of the same substance (TPP/Chit nanoparticles, chitosan and Nystatin); different lower-case letters in the lines indicate statistically significant differences (Mann-Whitney U test, $p < 0.05$) among substances at the same concentration. *Source* DeCarvalho et al., copyright © 2019 with permission from Elsevier B.V.

3.7 *Ionic Gelled Chitosan Nanocarriers for Effective Biofortification Process*

In this article, chitosan nanoparticle was generated by dropwise addition of chitosan in sodium tripolyphosphate (1% w/v) solution followed by stirred for 20 min. Then zinc sulphate was mixed with chitosan nanoparticle to obtain the zinc embedded chitosan nanoparticle, which was separated by centrifuged at 14,000 rotations per minute. The gradual increase of chitosan concentration (0.1–0.4 g%), and tripolyphosphate (0.5–2.0 g%) reflects the gradual increasement in particle size from 103 to 465 nm with average diameter of 200 nm. The optimal sized nanoparticle was developed from chitosan (0.3 g%), zinc sulphate 0.1 g percent and tripolyphosphate 1 g percent with 325 nm and +42.34 milivolt spherical nanoparticle and also observed with greater content of zinc in MACS 3125 and UC1114 wheat cultivars (Fig. 7).

This information stated the importance of zinc-containing chitosan nanoparticle for effective biofortification behavior [6].

3.8 *Ionic Gelled Chitosan Nanocarriers for Drug Delivery in Static Mixers*

Here chitosan nanoparticle was developed using static mixing process between chitosan and sodium tripolyphosphate solutions with a (5:1) ratio and (125:25) ratio flow rate followed by loading of salicyclic acid as model drug and cross filtered to obtain 50 ml drug-loaded nanosuspension. By this process, nanoparticle with 255 nm particle sizes was obtained but with the increasing flow rate of (250:50) to (375:75) the particle sizes were decreased up to 23 nm and 31 nm from initial. It was also

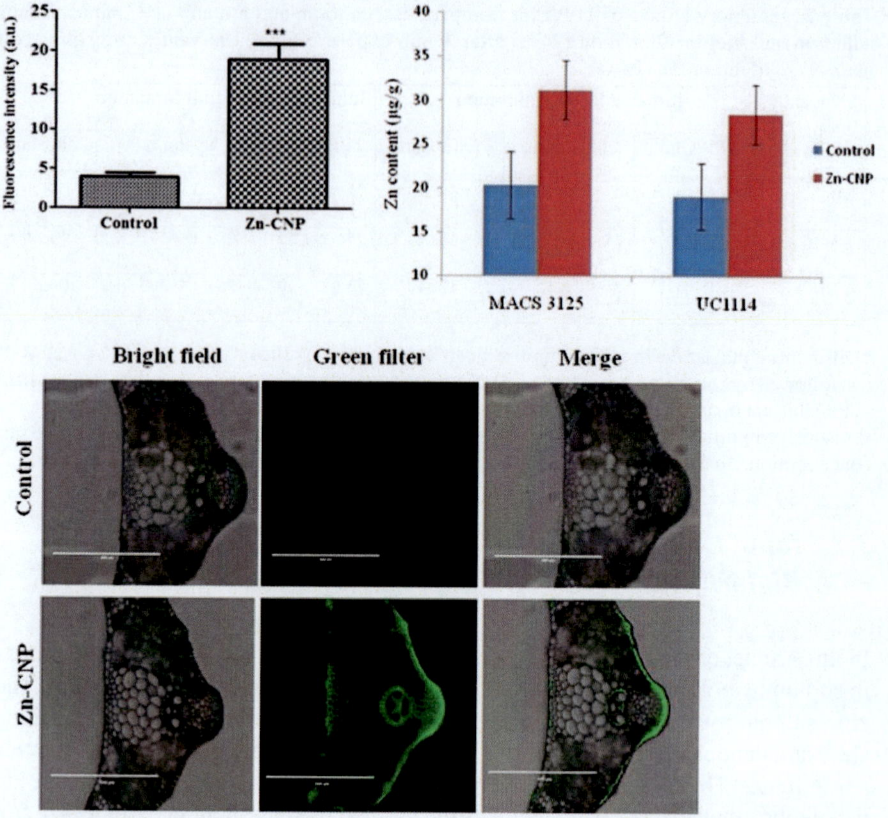

Fig. 7 **a** Fluorescence microscopy representative images for Zn localization in leaves using Zinquin dye (scale bar 400 μm), **b** Fluorescence intensity of Zinquin stain (Error bar shows mean ± SD, $n = 8$), $***p < 0.001$ (One-way ANOVA, Tuckey's multiple comparison test), **c** leaf Zn content (Error bar shows mean ± SD, $n = 3$). *Source* Deshpande et al. [6] with permission from Elsevier Ltd.

observed that step increase of chitosan concentration (0.25, 0.5, 0.75 and 1.0) mg/ml (Table 3), the particle size was increased up to two-fold (Fig. 8).

The average size of the particle was 200 nm with spherical shape with greater release of salicylic acid [7].

3.9 Ionic Gelled Chitosan Nanoparticle for Gene Delivery

In this article, nanoparticle was developed using chitosan (0.05–0.30) percent w/v and tripolyphosphate (0.7 ml) and finally obtained nanoparticle with (3:1, 4:1, 5:1, 6:1, 7:1) ratio between chitosan and crosslinker with ionic gelation method. The transmission electron microscopy revealed the particle size from 140 to 250 nm. The

Table 3 Effect of process parameters on the size and Polydispersity index of Chitosan Nanoparticle

Samples		Size (nm)	Polydispersity Index
Process Parameters	Variables		
Number of mixing elements	6	228	0.24
	12	232	0.201
Flow rate (ml/min) (Chitosan: tripolyphosphate)	18	217	0.210
	125:25	255	0.251
	250:50	232	0.201
	375:75	224	0.238
Chitosan Concentration (mg/ml)	500:100	239	0.246
	0.25	152	0.149
	0.50	232	0.201
Volume ratio (Chitosan: tripolyphosphate)	0.75	307	0.334
	1.0	377	0.459
	7:1	267	0.249
	5:1	232	0.201
	3:1	218	0.182

Source Dong et al., copyright © 2013 with permission from Elsevier Ltd

bovine serum albumin was used as protein model, which was developed using low molecular weight chitosan, crosslinker with (5:1) ratio and all the formulation was within 300–350 nm particle size range. The molecule became good delivery option of gene [8].

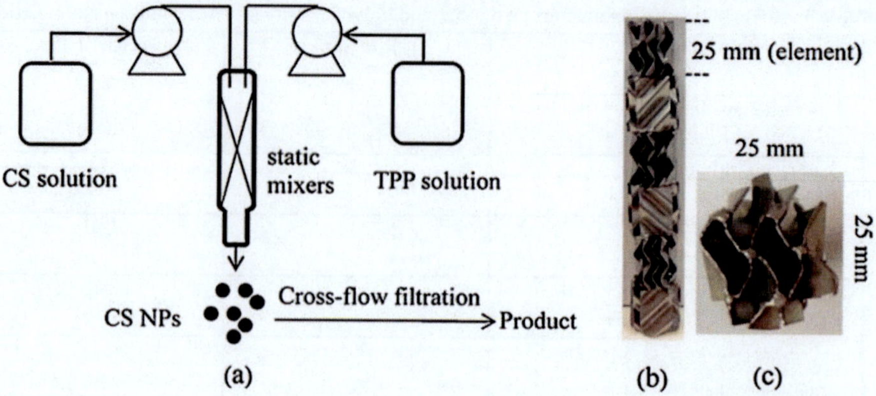

Fig. 8 Scheme of ionic gelation synthesis of CS NPs in static mixers (**a**), image of 1-segment static mixers composed of 6 elements (**b**) and top view of static mixers (**c**). *Source* Dong et al. [7] with permission from Elsevier Ltd.

3.10 Ionic Gelled Chitosan Nanoparticle for Delivery of Tacrine

In this article, nanoparticle was generated by the reaction between basic chitosan (1% weight/volume) with different concentrations (0.05, 0.1, 0.15, 0.2 and 0.25) percent weight/volume and 8 ml of acidic sodium tripolyphosphate(0.05, 0.1, 0.15, 0.2 and 0.25) percent weight/volume, followed by centrifuged between 500 and 1400 rotations per minutes for 15–45 min. Transmission electron microscopy and zetasizer outcomes observed with particle size with 90–100 nm and potential around +31 milivolt with 0.22 polydispersity index value. Zeta potential was increased up to 7 points after the encapsulation of positively charged tacrine. This study indicated the importance of ionic gelation for the delivery of tacrine [9].

3.11 Ionic Gelled Chitosan Nanoparticle for Delivery of Amoxicillin

Here, the nanoparticle was developed using mixing basic ureido conjugated chitosan (0.6, 1.2, 1.8, 2.4, 3.0 mg/mL concentration range in acetic acid)and sodium tripolyphosphate (concentration range 0.33, 0.67, 1.00, 1.33, 1.67 mg/mL) followed by maintaining ratios between chitosan derivative and crosslinker (2:1,3:1,4:1,5:1,6:1) and pH adjusted to obtain final nanosized particles. Here two different chitosan derivatives (such as type-1 contained with 12-Ureidododecanoic acid, chitosan, and type-2 contained with chitosan) were considered for delivery of amoxicillin. The particle size of drug loaded chitosan nanoparticle was 175.5 nm with 22.8 milivolt zeta potential, loading efficiency of 28.7 percent. But particle size was increased from 385.1 to 717.6 nm with ureido derived chitosan. Though chitosan contained nanoparticle showed greater stability and smaller particle size. Finally, release of amoxicillin from ureido chitosan nanoparticle was 80.1% whereas chitosan nanoparticle with 87.2%. These outcomes revealed the efficiency of ionic gelled chitosan nanoparticle for effective delivery of amoxicillin with greater inhibition of *H. pylori* [10].

3.12 Ionic Gelled Chitosan Nanoparticle for Effective Delivery of Enzyme

In this article, chitosan nanoparticle was generated upon reaction between basic chitosan (0.25%) m/v and sodium tripolyphosphate (0.2%) m/v followed by stirred at 800 rotations per minute and centrifuged to obtain a nanoparticle with (5:1) chitosan and crosslinker. The electron microscopy and polydispersity index were 784.4 nm & 761.6 nm and 0.207 & 0.104 without or with galactosidase enzyme, respectively; also zeta potential was minimized from 27.6 to 17.9, correlated with stability. The enzyme was 94% release from the formulation within 3H. This data indicated the potential of ionic gelled chitosan for delivery of enzyme [14].

3.13 Ionic Gelled Thiol Functionalized Chitosan Nanoparticle for Mercury Adsorption

Here, nanoparticle was developed upon reaction equimolar epichlorhydrin & cysteneaminium chloride and chitosan (4 g/L) to generate a thiol-functionalized chitosan followed by addition of sodium tripolyphosphate with a static flow rates (18, 20, 22, 24 ml/min). The particle size was minimized to 49 nm with flow rate increased from 18 to 24 ml/min of chitosan concentration. The average particle diameter was 48 nm with 0.19 polydispersity index. A peak at 2579 cm^{-1} was confirmed the presence of thiol group in formulation (Fig. 9).

This formulation was observed with maximum adsorption of mercury 50–250 mg/L concentration range [16].

3.14 Ionic Gelled Tripolyphosphate-Beta-Cyclodextrin-Chitosan Nanoparticle for Drug Delivery

In this article, nanoparticle was developed by the reaction between low molecular weight chitosan (2 g in 100 ml acetic acid solution), sodium tripolyphosphate (1.75,

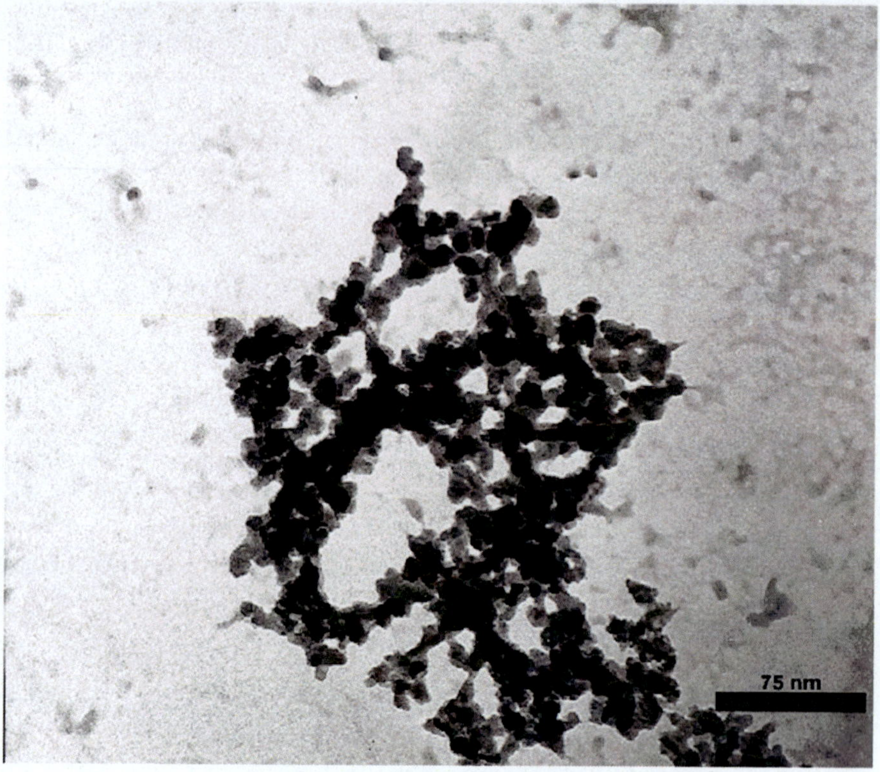

Fig. 9 The TEM image of thiol-functionalized CS NPs fabricated by using MF technique in the optimum conditions (scale bar of 75 nm). *Source* Nemati et al. [16] with permission from Elsevier Ltd.

2.1, and 2.6) mg/ml as crosslinking agent, followed by stirred at 300 rotations per minutes. Also, another chitosan nanoparticle was generated upon reaction between tripolyphosphate-beta-cyclodextrin mixture and chitosan (1.0, 1.5, 1.9, and 2.5) mg/ml with constant magnetically stirred at 1000 rotations per minute. As per X-ray diffraction data, peak near 26.87 and 15.12 confirmed the presence of tripolyphosphate and beta cyclodextrin, respectively. Particle size with 104.2 nm, polydispersity index of 0.346, and +27.33 milivolt of zeta potential were the characteristics of inclusion. This data indicated the importance of chitosan nano inclusion with cyclodextrin as drug delivery system [24].

3.15 Ionic Gelled Chitosan Nanoparticle with In Vitro Transfection Efficiency

In this article, nanoparticle was developed by the reaction with basic chitosan (5 mg/ml), sodium tripolyphosphate (2 mg/ml) in a reactor followed by stirred at 1000 rotations per minutes; also the nanoparticles were developed in different pH values (4.0, 5.0, 5.5). The enzyme luciferase PVAX1-Luc with concentration (330 μg/ml) was added to ionic gelled chitosan nanoparticle (5:8 ratio between chitosan and crosslinker) with DNA (10 and 20%) and chitosan ratios with different pH values 4.0, 5.0, 5.5; followed by vortexed for 1 min at 25 °C temperature. At pH value of 4.0, 5.0 and 5.5, particle size diameter, polydispersity index and zeta potential values were 142 nm, 190 nm, 0.6, 0.5, 44 milivolt, 43 milivolt; 211 nm, 312 nm, 0.53, 0.6, 35 milivolt, 32 milivolt; 249 nm, 255 nm, 0.4, 0.4, 37 milivolt, 36 milivolt, respectively. These nanoformulations were incorporated with HeLa cells. The greater luciferase activity was observed with chitosan nanoparticle with DNA (20%) with pH values 4.0 and 4.5. This data stated with the importance of chitosan nanoparticle for transfection efficiency [30].

3.16 Ionic Gelled Carboxyacyl Chitosan Nanoparticle for Effective Delivery of Taxane

In this article, nanoparticle was developed upon sonicated and centrifuged between carboxyacyl chitosan (0.5 g in acetic acid solution reacts with succinic or glutaric anhydride) and (0.25%) sodium tripolyphosphate at 7500 rotations per minutes for 40 min, followed by addition of 10 mg/ml of taxanes (paclitaxel and docetaxel) were added to chitosan nanoparticle and confirmed the formation of formulations by zeta potential, electron microscopy, and light scattering techniques. The particle size between 10–400 nm and 300–350 nm of apparent hydrodynamic diameter, [25–31] millivolt of zeta potential, [21–26] % of drug encapsulation efficiency and [6–13] % of drug loading efficiency have characterized the nanoparticles. The MTT assay revealed that cell viability was decreased with increasing concentration. The data indicated the effectivity of taxane delivery [31].

3.17 Ionic Gelled Chitosan Nanoparticle to Treat Human Skin Melanoma

Here, the nanoparticle was developed upon reaction between basic pH chitosan (1 mg/ml) (maintained at 5.0–6.0 with sodium hydroxide) with different tripolyphosphate and glucosamine ratios (0.05, 0.075, 0.1, 0.125, 0.150,0.175) followed by centrifugation at 750 rotations per minutes with the addition of bovine serum albumin

(0.2 mg/ml). The formulation was characterized by dynamic light scattering and encapsulation efficiency techniques. The particle size was 180–280 nm and zeta potential between +15 and +40 millivolt revealed the stability of the formulation. The degree of dissociation of chitosan was decreasing with increasing pH values, but polydispersity index remain unchanged. Also, bovine serum albumin entrapment efficiency was increased with crosslinker and glucosamine ratios. The nanoparticles were observed with greater efficiency against human skin cancer cell lines [32].

3.18 Ionic Gelled Chitosan Nanoparticle for Gene Delivery to Fish

Here, the nanoparticle was developed upon sonication between chitosan (3 mg/ml in acetic acid) and tripolyphosphate (1 mg/ml) at 16,000 rotations per minutes at 14 °C temperature for 30 min, followed by washed with different ratios of ethanol and lyophilized. Then the nanoparticle was centrifuged with plasmid DNA at 20,000 rotations per minutes at 10 °C temperature for 30 min. The particle sizes of the nanoparticles were 26.2 nm and 38.0 nm with spherical shape with 79.9% of encapsulation efficiency of DNA followed by stabilization determination through DNase-I and chitosanase enzymes. ELISA and RT-PCR analyses observed with greater efficacy with delivering gene [33].

3.19 Ionic Gelled Chitosan Microspheres with Vanillin/Tripolyphosphate for Protein Delivery

In this article, the formulation was developed using chitosan solution (10 mg/ml concentration) and bovine serum albumin (3.50 mg/ml), monovalent tetanus toxoid (1500 flocculation limit/ml), and divalent vaccine of diphtheria and tetanus toxoid (1500 flocculation limit/ml); these mixtures were considered as water phase. Another oil base was developed with liquid paraffin, span 80 and tween 80; then the emulsion was formed by constant stirring for 30 min. Emulsion was reacted with vanillin (12.5 mg/ml) and tripolyphosphate (12.5 mg/ml) to obtain the microspheres. As per the X-ray diffraction data, peak at 23.40; 43.79; 23.63, 43; 23.94, 44.01 were related with inversion of phase in different types of microspheres, also a distinct peak at 49 resembled with vanillin and crosslinking agent. The particle size of the microspheres was with 4 μm and zeta potential was within +47.7 to +66.2 milivolt range. The maximum yield was obtained with chitosan microsphere (86%) followed by other with albumin/tetanus/diphtheria with tetanus and maximum encapsulation efficiency and release were observed with chitosan-tetanus toxoid vaccine. This data correlated with the importance of protein delivery [34].

3.20 Reseveratrol Loaded Ionic Gelled Chitosan Nanoparticles for Cancer Therapy

In this article, nanoparticle was developed by reaction between basic chitosan solution (prepared by reaction between chitosan and poloxamer 188 in acetic acid solution) with ethanolic resveratrol (1 mg/ml or 5 mg/ml) solution and tripolyphosphate (1 mg/ml) solution. Another fluorescein-conjugated resveratrol loaded chitosan nanoparticle was developed using fluorescein isothiocyanate, resveratrol and chitosan. A peak at 880 cm^{-1} as per fourier transformed infrared spectroscopy has confirmed the presence of resveratrol and resveratrol loaded chitosan nanoparticle was observed with 172–217 nm of particle size with 30 milivolt zeta potential confirmed the stable dispersibility of microsphere. The release of resveratrol was quite greater in pH around 6.5 and also it maintained the antioxidative nature of the molecule with antiproliferative activity against hepatocellular carcinoma cell line (SMMC 721). This data clearly stated the applicability of the nanoformulation against hepatocellular carcinoma [35].

3.21 Interleukin Loaded Ionic Gelled Chitosan Nanoparticles for Colorectal Cancer Therapy

In this article, formulation was generated through reaction between chitosan, and tripolyphosphate with (2:1 or 4:1) ratio and various interleukin-12 solutions (0.5, 1.0, 1.5 and 2.0 mg/ml) was incorporated within the crosslinked chitosan (2 mg/ml) followed by incubation at room temperature for 30 min. The outcomes revealed that particle size diameter was within 178–372 nm with zeta potential was 24–53 millivolt range. The fluorescence intensity was minimized with particle between 100 and 200 nm in lung, kidney, liver, and spleen while it was higher with lower size particles. There was a significant observation in terms of interferon gamma was observed with animals treated with chitosan-interleukin-12 and ionically gelled chitosan-interleukin-12 (Fig. 10).

These data suggest the importance of ionic gelation process to deliver interleukin-12 embedded with chitosan nanoparticle [36].

3.22 Ionically Gelled Chitosan Nanoparticle with Wound Healing Activity

In this paper, formulation was developed upon reaction between chitosan and tripolyphosphate with (4:1) ratio, and the mixture was centrifuged at 16,000 rotations per minute for total 110 min, followed by desiccation. Then the formulation was fed into animals and compared the wound healing property after comparing with

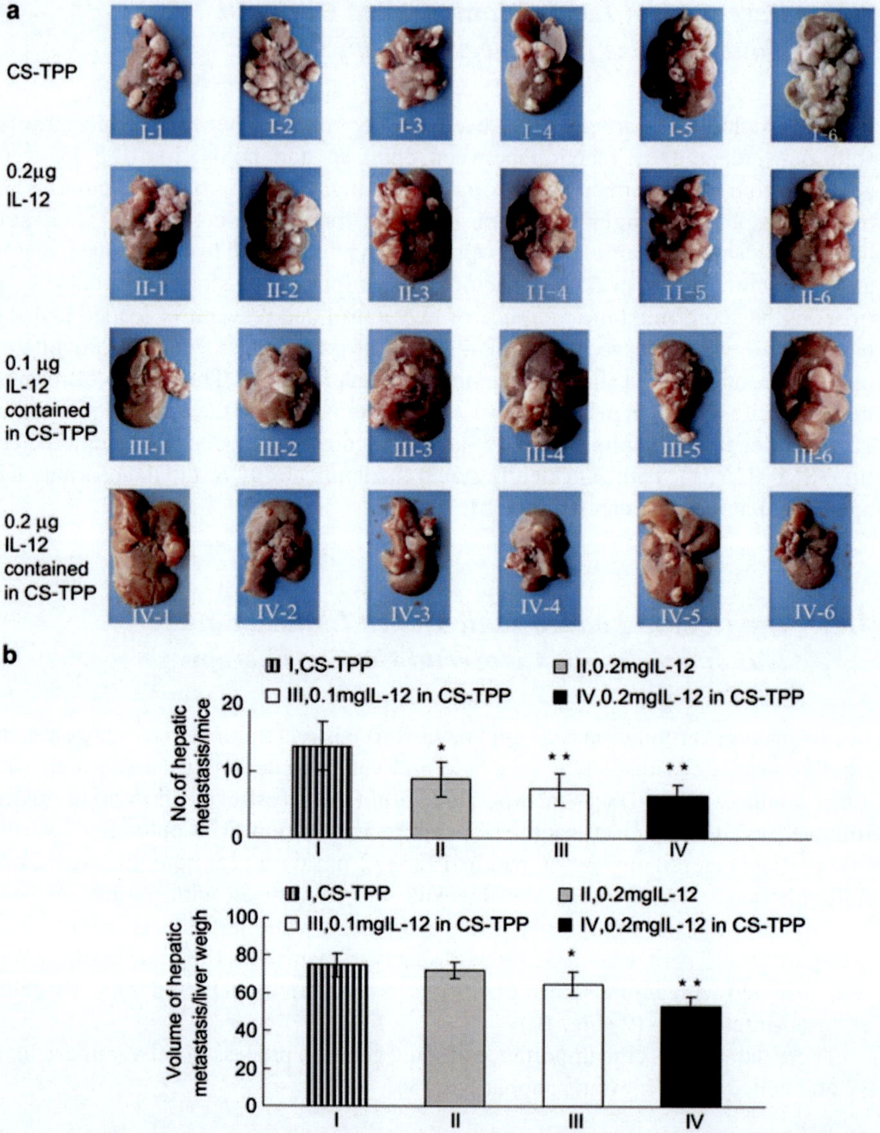

Fig. 10 Prevention of CRC hepatic metastasis by intravenous delivery of CS-TPP/IL-12 (200 nm). BALB/c mice received a spleen injection of CT26 carcinoma cells to induce hepaticmetastases. Two weeks later, each mouse was injected intravenously on a daily basis with CS-TPP nanoparticles alone (group I), 0.2 mg IL-12 (group II), 0.1 mg IL-12 contained in CS-TPP (group III), and 0.2 mg IL-12 contained in CS-TPP (group IV). The mice were sacrificed 21 days after tumor challenge, the number of hepatic metastases counted and volume of metastasis measured. **a** Representative photographs of hepatic metastases from each group; **b** Effect of CS-TPP/IL-12 on number of hepatic metastases and volume of hepaticmetastasis. Data are presented as mean ± SD, n ¼ 6, *$p < 0.05$, **$p < 0.01$, versus CS-TPP group. *Source* Xu et al. [36] with permission from Elsevier Ltd.

control followed by histopathology of their anterior lobes of abdomen. The outcomes revealed that mean patrtcile size was 60.2 nm with three characteristic peaks at 11.48, 11.84, 23.58 as per X-ray diffraction confirmed the formation of the formulation. The histopathology suggested that a yellow structure was developed and new granulation were observed around the wound within first fortnight of treatment. These data clearly stated the importance of chitosan ionic gelled nanoparticle for the treatment of wound [37].

4 Conclusion

The chapter provides a brief overview on chitosan micropartcle or nanoparticle generated through ionic gelation technique using sodium tripolyphosphate as crosslinking agent that have been used for drug delivery of MDA-MB-231, doxorubicin to treat breast/skin/colorectal cancer, bovine serum albumin, *Saturejahortensis* essential oil to treat acaricide, of zinc in MACS 3125 and UC1114 wheat cultivars, particular gene, tacrine, amoxicillin to treat *Helicobacter pylori*, luciferase PVAX1-Luc with in vitro transfection efficiency and also these nanoparticles showed inhibition of *Aedes aegypti, Candida albicans* with wound healing and mercury adsorption properties. However, this technique to develop this micro/nanoparticle observed with higher surface area, polydispersity index, and greater topological index. Always scientist follows the natural and ancient source to deliver a drug with naturalistic biocompatible and biodegradable way. The future scope of this chapter shall focus on the investigation of natural polymers with its own activity profile which can synergize the activity of Active Pharmaceutical Ingredient.

References

1. Abdelrahman MA, Hudson SM (2019) Chitosan nanoparticles: Polyphosphates cross-linking and protein delivery properties. Int J Biological Macromol 136:133–142
2. Ahmadi Z, Saber M, Akbari A, Mahdavinia GR (2018) Encapsulation of Saturejahortensis L. (Lamiaceae) in chitosan/TPP nanoparticles with enhanced acaricide activity against Tetranychusurticae Koch (Acari: Tetranychidae). Ecotoxicol Environment Safety 161:111–119
3. Anand M, Sathyapriya P, Maruthupandy M, Beevi AH (2018) Synthesis of chitosan nanoparticles by TPP and their potential mosquito larvicidal application. Front Lab Med 2(2):72–78
4. Anandhakumar S, Krishnamoorthy G, Ramkumar KM, Raichur AM (2017) Preparation of collagen peptide functionalized chitosan nanoparticles by ionic gelation method: An effective carrier system for encapsulation and release of doxorubicin for cancer drug delivery. Material Sci Eng C 70:378–385
5. DeCarvalho FG, Magalhaes TC, Teixeira NM, Gondim BLC, Carlo HL, Santos RL, Oliveira AR, Denadai AML (2019) Synthesis and characterization of TPP/chitosan nanoparticles: Colloidal mechanism of reaction and antifungal effect on C. albicans biofilm formation. Material Sci Eng C 104:109885

6. Deshpande P, Dapkekar A, Oak MD, Paknikar KM, Rajwade JM (2017) Zinc complexed chitosan/TPP nanoparticles: A promising micronutrient nanocarrier suited for foliar application. CarbohydrPolym 165:394–401
7. Dong Y, Ng WK, Shen S, Kim S, Tan RBH (2013) Scalable ionic gelation synthesis of chitosan nanoparticles for drug delivery in static mixers. CarbohydrPolym 94:940–5
8. Gan Q, Wang T, Cochrane C, McCarron P (2005) Modulation of surface charge, particle size and morphological properties of chitosan–TPP nanoparticles intended for gene delivery. Colloids Surfaces B: Biointerface 44:65–73
9. Hassani S, Laouini A, Fessi H, Charcosset C (2015) Preparation of chitosan-TPP nanoparticles using microengineered membranes-Effect of parameters and encapsulation of tacrine. Colloids Surf A: Physico chem Eng Aspects 482:34–43
10. Jing ZW, Jia YY, Wan N, Luo M, Huan ML, Kang TB, Zhou SY, Zhang BL (2016) Design and evaluation of novel pH-sensitive ureido-conjugated chitosan/TPP nanoparticles targeted to Helicobacter pylori. Biomaterial 84:276–285
11. Koukaras EN, Papadimitriou SA, Bikiaris DN, Froudakis GE (2012) Insight on the Formation of Chitosan Nanoparticles through Ionotropic Gelation with Tripolyphosphate. | Mol. Pharmaceutics 9:2856–2862
12. Kunjachan S, Jose S, Lammers T (2010) Understanding the mechanism of ionic gelation for synthesis of chitosan nanoparticles using qualitative techniques. Asian J Pharmaceutics 4:148–153. https://doi.org/10.4103/0973-8398.68467
13. Lee E, Park SJ, Lee JH, Kim MS, Kim CH (2016) Preparation of chitosan–TPP nanoparticles and their physical and biological properties. Asian J Pharmaceutical Sci 11:166–167
14. Leichner C, Jelkmann M, Prufert F, Laffleur F, Schnurch AB (2019) Intestinal enzyme delivery: Chitosan/tripolyphosphate nanoparticles providing a targeted release behind the mucus gel barrier. European J Pharmaceutics Biopharmaceutics 144:125–131
15. Mohamed EIB, Entsar IR (2011) A Biopolymer Chitosan and Its Derivatives as Promising Antimicrobial Agents against Plant Pathogens and Their Applications in Crop Protection. Int J Carbohydr Chem 2011:1–29
16. Nemati Y, Zahedi P, Baghdadi M, Ramezani S (2019) Microfluidics combined with ionic gelation method for production of nanoparticles based on thiol-functionalized chitosan to adsorb Hg (II) from aqueous solutions. J Environmental Management 238:166–177
17. Pal D, Kumar S, Saha S (2017) Antihyperglycemic activity of phenyl and ortho-hydroxy phenyl linked imidazolyl triazolo hydroxamic acid derivatives. Int J Pharmacy Pharmaceutical Sci 9(12):247–251
18. Pal D, Nayak AK, Hasnain MS, Saha S (2019) Pharmaceutical Applications of Chondroitin. Volume 3: Animal-Derived Polymers, Natural Polymers for Pharmaceutical Applications. Apple Academic Press, CRC Press (Taylor and Francis Group)
19. Pal D, Nayak AK, Saha S (2018) Interpenetrating Polymer Network Hydrogels of Chitosan: Applications in Controlling Drug Release. Cellulose-Based Superabsorbent Hydrogels. Polymers and Polymeric Composites: A Reference Series. Springer, Cham. Springer International Publishing AG, part of Springer Nature. Polymers and Polymeric Composites: A Reference Series, https://doi.org/10.1007/978-3-319-76573-0_57-1
20. Pal D, Nayak AK, SupriyoSaha S (2019) Cellulose Based Hydrogel. Springer Nature Singapore Pte Ltd. 2019. Natural Bio-active Compounds 285–332, https://doi.org/10.1007/978-981-13-7154-7_10
21. Pal D, Saha S (2019) Chondroitin: a natural biomarker with immense biomedical applications. RSC Advances 9(48):28061–28077
22. Pal D, Saha S (2019) Current Status and Prospects of Chitosan-Metal Nanoparticles and Their Applications as Nanotheranostic Agents. Nanotheranostics, Springer Nature Switzerland AG, pp 79–114. https://doi.org/10.1007/978-3-030-29768-8_5
23. Pal D, Saha S, Nayak AK, Hasnain MS (2019) Marine-Derived Polysaccharides: Pharmaceutical Applications. Volume 2: Marine- and Microbiologically Derived Polymers, Natural Polymers for Pharmaceutical Applications. Apple Academic Press, CRC Press (Taylor and Francis Group)

24. Pant A, Negi JS (2018) Novel controlled ionic gelation strategy for chitosan nanoparticles preparation using TPP-β-CD inclusion complex. European J Pharmaceutical Sci 112:180–185
25. Saha S, Pal D (2017) Encyclopedia of Physical Organic Chemistry. Wiley Interscience, NewYork, pp 1–22
26. Saha S, Pal D, Kumar S (2016) Design, synthesis and antiproliferative activity of hydroxyacetamide derivatives against HeLa cervical carcinoma cell and breast cancer cell line. Tropical J Pharmaceutical Res 15(7):1319–1326
27. Saha S, Pal D, Kumar S (2017) Antifungal and Antibacterial Activities of Phenyl and Ortho-Hydroxy Phenyl Linked Imidazolyl Triazolo Hydroxamic Acid Derivatives. Inventi Rapid: Med Chem 2:42–49
28. Saha S, Pal D, Kumar S (2017) Hydroxyacetamide derivatives: cytotoxicity, genotoxicity, antioxidative and metal chelating studies. Indian J Experiment Biol 55:831–837
29. Shweta A, Ankita L, Aakriti T, Vijay K, Imran M, Anita KV (2015) Versatality of Chitosan: A Short Review. J Pharm Res 4(3):125–134
30. Sipoli CC, Santana N, Shimojo AAM, Azzoni A, Torre LG (2015) Scalable production of highly concentrated chitosan/TPP nanoparticles in different pHs and evaluation of the in vitro transfection efficiency. Biochemic Eng J 94:65–73
31. Skorik YA, Golyshev AA, Kritchenkov AS, Gasilova ER, Poshina DN, Sivaram AJ, Jayakumar R (2017) Development of drug delivery systems for taxanes using ionic gelation of carboxyacyl derivatives of chitosan. Carbohydr Polym 162:49–55
32. Stie MB, Thoke HS, Issinger OG, Hochscherf J, Guerra B, Olsen LF (2019) Delivery of proteins encapsulated in chitosan-tripolyphosphate nanoparticles to human skin melanoma cells. Colloid Surfaces B: Biointerface 174:216–223
33. Vimal S, Taju G, Nambi KSN, Majeed SA, Babu VS, Ravi M, Hameed ASS (2012) Synthesis and characterization of CS/TPP nanoparticles for oral delivery of gene in fish. Aquaculture 358–359:14–22
34. Walke S, Srivastava G, Nikalje M, Doshi J, Kumar R, Ravetkar S, Doshi P (2015) Fabrication of chitosan microspheres using vanillin/TPP dual crosslinkers for protein antigens encapsulation. CarbohydrPolym 128:188–198
35. Wu J, Wang Y, Yang H, Liu X, Lu Z (2017) Preparation and biological activity studies of resveratrol loaded ionically cross-linked chitosan-TPP nanoparticles. CarbohydrPolym 175:170–177
36. Xu Q, Guo L, Gu X, Zhang B, Hu X, Zhang J, Chen J, Wang Y, Chen C, Gao B, Kuang Y, Wang S (2012) Prevention of colorectal cancer liver metastasis by exploiting liver immunity via chitosan-TPP/nanoparticles formulated with IL-12. Biomaterial 33:3909–3918
37. Zhao A, Wang T, Yao M, Li H (2011) Effects of chitosan-TPP nanoparticles on hepatic tissue after severe bleeding. J Med College PLA 26:283–292

Chapter 5
Ionically Gelled Carboxymethyl Polysaccharides for Drug Delivery

Mohsen Khodadadi Yazdi, Mohammad Reza Ganjali, Payam Zarrintaj, Babak Bagheri, Yeu Chun Kim. and Mohammad Reza Saeb

Abstract Polysaccharides are building blocks of many drug delivery systems with outstanding performance. Carboxymethyl derivatives obtained from polysaccharides improve solubility and physiochemical properties making them good alternatives for various applications. Carboxymethyl polysaccharides bearing cationic charges can be physically crosslinked via electrostatic and ionic interaction with polycationic species, small molecules, and metal ions to induce integrity. In this chapter, drug delivery platforms based on carboxymethyl polysaccharides in which ionic species are used as crosslinking agents which are outlined. We highlight that ionically crosslinked carboxymethyl polysaccharides carriers are promising candidates for delivery of various drugs such as chemotherapeutics, anti-inflammatory drugs, and protein-based therapeutic agents.

Keywords Carboxymethyl polysaccharide · Ionic crosslinking · Drug delivery · Controlled release

---

M. K. Yazdi · M. R. Ganjali
Center of Excellence in Electrochemistry, School of Chemistry, College of Science, University of Tehran, Tehran, Iran

M. R. Ganjali (✉)
Biosensor Research Center, Endocrinology and Metabolism Molecular-Cellular Sciences Institute, Tehran University of Medical Sciences, Tehran, Iran
e-mail: ganjali@ut.ac.ir

P. Zarrintaj
School of Chemical Engineering, Oklahoma State University, 420 Engineering North, Stillwater, OK 74078, USA

B. Bagheri · Y. C. Kim
Department of Chemical and Biomolecular Engineering, Korea Advanced Institute of Science and Technology (KAIST), 291 Daehak-ro, Yuseong-gu, Daejeon 34141, Republic of Korea

M. R. Saeb
Université de Lorraine, CentraleSupélec, LMOPS, F-57000 Metz, France
e-mail: mrsaeb2008@gmail.com

A. K. Nayak et al. (eds.), *Ionically Gelled Biopolysaccharide Based Systems in Drug Delivery*, Gels Horizons: From Science to Smart Materials,
https://doi.org/10.1007/978-981-16-2271-7_5

1 Introduction

The concept of drug delivery has emerged since early 1950s [1]. Since then, various materials have been utilized to make drug delivery systems (DDSs). However, polymers have attracted much interests compared to other biomaterials [2]. Both natural and synthetic polymers are among the most utilized materials in DSs. More recently, theranostic platforms, which combine both diagnosis and therapy in one system, have been introduced which are highly interesting for cancer treatment [3].

Synthetic polymers, due to versatile chemistry, reproducibility, adjustable properties, have attracted tremendous interests in tissue engineering and drug delivery [4, 5]. Degradable synthetic polymers are majorly based on aliphatic polyesters such as poly(lactic acid) (PLA), poly(glycolic acid) (PGA), poly (lactic-co-glycolic acid) (PLGA), poly(ε-caprolactone) (PCL), and poly(decalactone) [2]. PLGA is the most well-known biodegradable polyester which have been widely utilized in drug delivery applications [6]. Degradation time of PLGA could be adjusted based on molecular weight and the ratio of lactic acid (LA) to glycolic acid (GA).On the other hand, synthetic non-degradable polymers such as ethylene vinyl acetate copolymers (EVA), polyvinyl alcohol (PVA), polyvinylpyrrolidone (PVP), and polyurethane (PU) have also been used to fabricate DDSs [7, 8].

Natural polymers such as polysaccharides (e.g. agarose and chitosan) and proteins (e.g. collagen and silk fibroin) benefit from biocompatibility, low cost, tissue-like structure, hydrophilicity, and biodegradation. These abundant polymers constitute different parts of plant and animal bodies [9]. Proteins are abundant in animal bodies serving as enzymes, hormones, antibodies, receptors, and structural species. On the other hand, polysaccharides are found in both animal and plants. Starch, glycogen, and cellulose are the most abundant polysaccharides in nature, which serve as food storage (starch in plants and glycogen in animals) and structural components (in plants). These polysaccharides are the product of a natural carbon dioxide capturing process utilizing photosynthesis and biosynthesis processes. Polysaccharides are polymer that is composed of monosaccharide units; simple sugars such as glucose, fructose, and galactose are monosaccharides that constitute main building blocks of various polysaccharides [10].

On the other hand, there are several natural polysaccharides extracted from some species such as seaweeds and crab shells which not as abundant as starch, glycogen, and cellulose. The chemical structure of these polysaccharides is different from abundant polysaccharides; indeed, some of hydroxyl groups of abundant polysaccharides are substituted with carboxyl (e.g. in pectin and agarose), amino (e.g., chitosan), or other chemical functionalities [11]. Some of the polysaccharides only have hydroxyl as pendant groups (e.g. cellulose, dextran, guar gum, and pullulan) in their chemical structure, others contains uronic acids (e.g. alginate and pectin) or nitrogen containing functional groups (e.g. chitosan and chitin) [12]. These functional groups significantly affect the solubility and physiochemical properties of various polysaccharides.

Beside natural-based biosynthesis processes, polysaccharides can be artificially modified using various physical, chemical, and biological processes to endow them with novel physical/chemical properties and bioactivities [13]. Physical and biological modifications can only alter the molecular weight of polysaccharides while versatile chemical modifications can significantly affect the physiochemical behavior of the resulting modified polysaccharides. Acetylation, sulfation, carboxymethylation, and alkylation are well-known chemical modification strategies used for polysaccharides. Carboxymethylation is a chemical modification technique that introduces carboxymethyl groups to polymer chain in order to increase polysaccharides solubility. Carboxymethylation can be used for various polysaccharides such as cellulose, chitin, and dextran [14, 15].

Carboxymethylation usually significantly improves solubility of polysaccharides; indeed, carboxymethyl cellulose (CMC) is highly water-soluble while cellulose is not soluble in water and organic solvents. Furthermore, carboxymethyl functional groups bearing negative charges can form polyelectrolyte complexes (PEC) with positively charged polymers such as chitosan (the only positively charged natural polysaccharide) [16–18]. Electrostatic interactions between polymer chains bearing opposite charges result in complexation. Moreover, small molecules or metal ions could also interact electrostatically with negatively charged polymers. Such electrostatic interactions can be utilized to prepare polymer nanoparticles that can serve as drug carriers or theranostic platforms [19]. Thesemethods are subgroups of physical crosslinking strategies that are discussed in the next section.

2 Crosslinking Methods

Crosslinking indicates linking two or more polymer chains together via covalent or physical bonds. This process usually includes utilization of crosslinking agents. Chemical crosslinkers have been widely utilized in polymer industries. Crosslinking of thermosetting resin and rubbers is known as curing and vulcanization, which is carried out through chemical reactions. Types of crosslinkers and their density highly affect the physiochemical properties of the obtained materials [20]. Most of the chemical crosslinking reactions (based on different reactions such as click chemistry, Diels-Alder reaction, Schiff base reactions, and Michael additions) make robust but irreversible bonds between adjacent polymer chains [21]. On the other hand, chemical crosslinking in the biological systems, such as formation of disulfide and isopeptide bonds, play key roles in making mechanically robust structures such as cartilage, skin, and hair [22]. Enzymes catalyze many of such crosslinking reactions in biological systems.

Physical crosslinking results in weaker and reversible interactions between polymer chains. For example, divalent metals such as calcium ions physically crosslink sodium alginate to make beads, hydrogels, and fibers [23, 24]. However, it is possible to remove ions and disassemble these ionic crosslinked structures through chelating agents such as sodium citrate and ethylenediaminetetraacetic acid

(EDTA). Furthermore, borax can crosslink polyvinyl alcohol (PVA) through physical hydrogen bonds [25]. Hydrogen bonding play pivotal role in DNA structure, as well. Besides, these physical crosslinking processes play key role in many natural derived polymers such as agarose, collagen, and so on. Normally, physical crosslinking is occurred between two species (two polymer chains or polymer chain with small molecules); however, recently, the concept of intramolecular crosslinking has also emerged to make single-chain nanoparticle (SCNP) [26].

Polymer hydrogels, which mimic the mechanical behavior of tissue, have been widely applied in various tissue engineering and drug delivery systems [27, 28]. In polymer, various physical and chemical crosslinking strategies are used, as represented in Fig. 1 [21]. These strategies can be used to make DDSs, as well. Physical crosslinking is based on ionic or electrostatic interactions, hydrogen bonding, hydrophobic interactions, crystallization, metal chelation, and host–guest interactions [21]. Dynamic ionic crosslinks could bestow elastomers with robustness and extraordinary stretchability [29]. Ionic crosslinking denotes, crosslinking using small molecules [30]. On the other hand, coacervation is the electrostatic interactions between oppositely charged polymer chains, and ionic and electrostatic interactions are among most utilized physical crosslinking methods.

Carboxymethyl bears anionic charge; thus, carboxymethyl polysaccharides are anionic polymers. It is possible to crosslink these polymers with metal cations, small molecule cations, and polycationicpolymers such as chitosan, polyethyleneimine (PEI), and poly-L-lysine (PLL) [31].

3 Carboxymethylation

Some of the polysaccharides such as cellulose are poorly soluble in water and organic solvents. In this regard, chemical modifications have been used to increase their solubility. Carboxymethylation is a robust strategy to make water-soluble polysaccharide derivatives, especially carboxymethyl cellulose (CMC) [32]. CMC benefits from biocompatibility, bioadhesive properties, and biodegradability [33]. Furthermore, carboxymethylation has been utilized to make carboxymethyl derivatives of other polysaccharides, as well [34, 35]. Multivalent metal cations and polycationic species can be used to crosslink carboxymethyl polysaccharides producing hydrogels. For example, composite hydrogels based on CMC/bentonite were utilized for controlled release of herbicide in which Fe^{3+} ions were used to crosslink CMC [36].

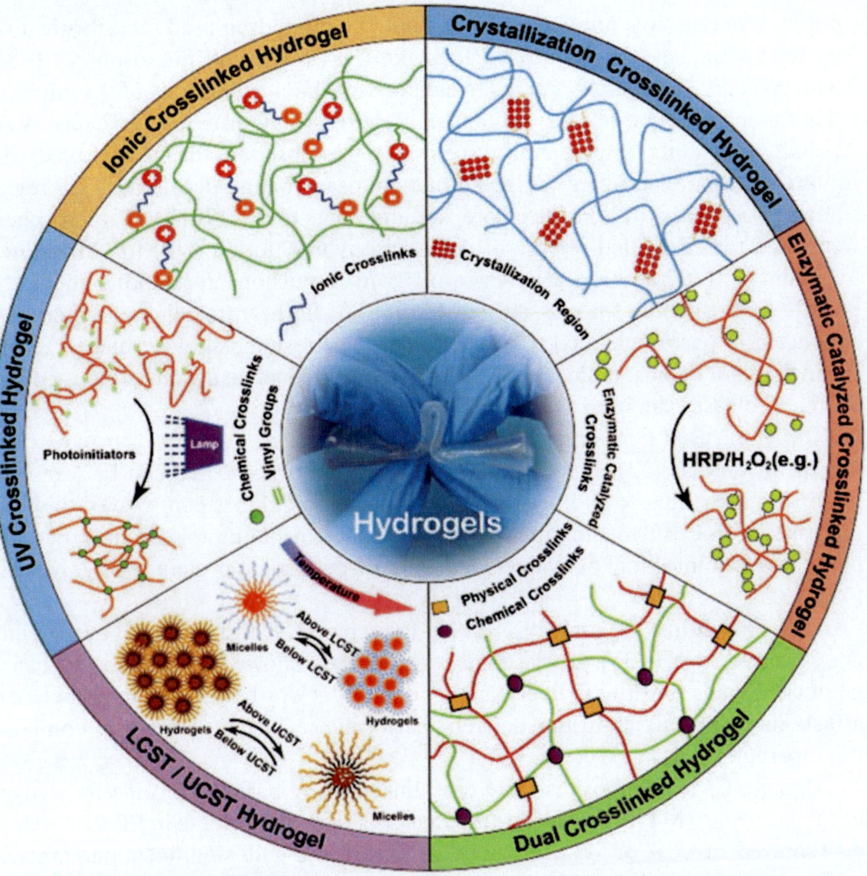

Fig. 1 Most commonly utilized crosslinking strategies to make polymers hydrogel [21]

4 Carboxymethyl Polysaccharides in Drug Delivery

4.1 Carboxymethyl Cellulose (CMC)

CMC and its blends with other water-soluble polymers can be utilized to make delivery vehicles for various therapeutic agents [37–39]. Various metal ions such as Fe^{3+} can be utilized as ionic crosslinking agent for CMC to improve its integrity. However, Fe^{3+}-crosslinked CMC carriers fail to withstand acidic conditions (e.g., gastric pH) limiting the oral administration of CMC-based DDSs. Blending and coating with other polymers may compensate for this limitation [40]. Bulutused sodium CMC (NaCMC)/PVA blends and coated them with chitosan to make flurbiprofen (FBP)-loaded microsphere for controlled release applications [41]. FBP is a poorly water-soluble non-steroidal anti-inflammatory drug used for arthritic disorders treatment. Fe^{3+}chelation used to make (NaCMC)/PVA microspheres (~3.8–9 μm) and chitosan coating applied to enhance mechanical integrity of the microparticles and inhibit burst release of the FBP. Relativelyhigh drug loading capacity (ca. 22–40%, depending on preparation methods), was observed for the microparticles. In vitro release, experiments revealed that chitosan coating decelerate FBP release from microspheres [41]. Furthermore, sodium alginate (SA)/NaCMC microspheres were used for controlled release of donepezil hydrochloride (DP) for Alzheimer's treatment [42]. They used a water-in-oil (w/o) emulsion crosslinking method to prepare microspheres for oral administration of DP. In vitro release, experiments conducted in gastrointestinal tract pH values. The resulted showed that increasing SA in polymer blends, increase DP release rate while increasing crosslinking degree decrease release rate.

Kim et al. used Fe^{3+}-crosslinked SA/CMC beads for delivery of protein-based therapeutic agents [43]. Electrostatic interactions between CMC and Fe^{3+} result in polymer chain collapse and bead formation. They found that pore size of the beads increased with CMC loading in the hydrogel beads. Albumin release from hydrogel beads in gastrointestinal pH values revealed pH-dependent protein release behavior [43].

$FeCl_3$ was utilized as ionic crosslinker to prepare NaCMC hydrogel beads which are sensitive to pH [33]. These nanoporous beads showed good stability can be good candidates for drug delivery applications. It is worth mentioning that chelator affects the synthesis and drug delivery performance of chitosan-CMC composite microparticles [44].

Alginate-CMC composites were crosslinked using barium chloride for delivery of methotrexate (MTX) for cancer treatment [45]. In vitro release in PBS (37 °C, pH 7.4) showed ca. 98% of MTX released just during 5 h with significant burst release.

Dey et al. utilize Al^{3+} cations to carboxymethyl locust bean gum (CMLBG)/alginate hydrogels for oral delivery of glipizide, an anti-diabetic medication [46]. High loading capacity (93-98%) was observed for glipizide. Besides, in vivo experiment on rats showed remarkable hypoglycemic activity.

4.2 *Carboxymethyl Chitosan/Chitin (CMCS/CMCN)*

Chitosan is the only natural cationic polysaccharide; it can be crosslinked via chemical agents such as glutaraldehyde or physical crosslinking through anions and anionic molecules (e.g. tripolyphosphate, TPP) or polyanions (e.g., polyphosphates) [47, 48]. Contrary to chitosan that bears positively charges, carboxymethyl chitosan (CMCS) is negatively charged. Hence, cationic species such as multivalent metal ions can induce ionic crosslinking in CMCS. Ca^{2+} ions play key role in crosslinking CMCSs and nanoparticles formation [49]. These pH-sensitive nanoparticles can be utilized in making DDSs. On the other hand, Lu et al. embedded Cu nanoparticles into alginate-CMCS mixture [50]; gradual release of Cu^{2+} ions from nanoparticles, results in ionic crosslinking of these polyanions; beside, alginate-CMCS/Cu scaffolds showed enhanced antibacterial effect and improved osteogenesis which are interesting in bone tissue engineering. Indeed, controlled release of metal cations not only induce crosslinking but also it shows antibacterial effects.

Mi et al. used Fe^{+3} cations to crosslink carboxymethylchitin (CMCN) to prepare microspheres [51]. The obtained microspheres were applied as pH-sensitive delivery vehicles for 6-mercaptopurine.

4.3 Other Polysaccharides

Chemical modification of starch provides various cheap biopolymers for diverse range of applications [52]. Carboxymethyl starch (CMST) has been used as biomaterial for making drug delivery or theranostic platforms [53–55]. Many researcher used the sodium trimetaphosphate (STMP) as chemical crosslinker to prepare CMST microgels [56, 57]. In most of research experiments, complex coacervation (i.e., electrostatic interactions between two oppositely charged ionic polymers) was utilized as physical crosslinking strategy to make drug nanocarriers based on CMST [58, 59]. It seems that no metal ions have been utilized to crosslink CMST for drug delivery applications, so far.

Carboxymethyl dextran (CMD) has been used to make various drug delivery platforms [60, 61]. This chemically modified polysaccharide has several functional groups such as hydroxyl, carboxyl, and aldehyde which can interact with drug molecules, other polymers, and ionic species [62]. Loaded drug can be released through polymer degradation [61]. In the most of the research studies, cationic polymers, especially chitosan, have been used to electrostatically crosslink CMD and making drug delivery platforms. For example, Lin et al. used chitosan to coacervate CMD and prepare 5-fluorouracil (5-FU)- or indocyanine green (ICG)-loaded chitosan-CMD nanoparticles [63].

Hyaluronic acid (HA) is an important naturally occurring polysaccharide benefiting from biocompatibility and biodegradation, in vivo [64]. Both HA and its derivatives have been widely used in biomedical applications [65]. Specially, HA-based materials have been widely applied in manufacturing targeted drug delivery platforms for cancer therapy [66, 67]. Carboxymethyl hyaluronic acid (CMHA) can be utilized to coat and stabilized calcium phosphate nanoparticles [68]. Calcium phosphate has been widely used in development of DDSs [69]. Chemotherapeutic agents-loaded CMHA-coated calcium phosphate nanoparticles effectively target human breast cancer cells and release the loaded drug [68].

Carboxymethyl guar gum (CMGG) beads were utilized for delivery of bovine serum albumin (BSA, model drug) through gastrointestinal tract [70]. Divalent barium (Ba^{2+}) and calcium (Ca^{2+}) were used as ionic crosslinking agents for CMGG. Minimal drug release was observed in simulated gastric pH during 6 h while nearly, all encapsulated drug was released in simulated intestinal fluid (SIF) [70].

Carboxymethyl xanthan gum (CMXG) was utilized to make drug carriers [71]. In this systems, $CaCl_2$ solution was utilized to crosslink CMXG.

5 Conclusion

Polysaccharides and their derivative are natural/green polymers widely used for tissue engineering and drug delivery applications. Carboxymethyl polysaccharides are important groups of chemically-modified polysaccharides bearing negative charges. Hence, electrostatic and ionic interactions resulting from such structures can induce physical cross-links for biomedical applications. Ionically gelled carboxymethyl polysaccharides can be utilized as drug delivery systems for diverse therapeutic agents. They can preserve drug bioactivity in oral administration; besides, they can play role in targeted drug delivery to tumor.

References

1. Park K (2014) Controlled drug delivery systems: past forward and future back. J Controlled Release 190:3–8
2. Rahmati M, Mills SK, Urbanska AM, Saeb MR, Venugopal JR, Ramakrishna S, Mozafari M (2021) Electrospinning for tissue engineering applications. Prog. Mater. Sci. 117:100721
3. Sharma R, Mody N, Agrawal U, Vyas S (2017) Theranostic nanomedicine; a next generation platform for cancer diagnosis and therapy. Mini Rev Med Chem 17(18):1746–1757
4. Shokri Z, Seidi F, Karami S, Li C, Saeb MR, Xiao H (2021) Laccase immobilization onto natural polysaccharides for biosensing and biodegradation. Carbohyd Polym 262:117963
5. Li J, Mooney DJ (2016) Designing hydrogels for controlled drug delivery. Nature Rev Mat 1(12):16071
6. Mir M, Ahmed N, Rehman A (2017) Recent applications of PLGA based nanostructures in drug delivery. Colloids and Surf B: Biointerfaces 159:217–231
7. Schneider C, Langer R, Loveday D, Hair D (2017) Applications of ethylene vinyl acetate copolymers (EVA) in drug delivery systems. J Controlled Release 262:284–295
8. Jensen BE, Dávila I, Zelikin AN (2016) Poly (vinyl alcohol) physical hydrogels: matrix-mediated drug delivery using spontaneously eroding substrate. J Phys Chem B 120(26):5916–5926
9. Doppalapudi S, Katiyar S, Domb AJ, Khan W (2015) Biodegradable natural polymers. In: Advanced polymers in medicine. Springer, pp 33–66
10. Khosravi A, Fereidoon A, Khorasani MM, Naderi G, Ganjali MR, Zarrintaj P, Saeb MR, Gutiérrez TJ (2020) Soft and hard sections from cellulose-reinforced poly (lactic acid)-based food packaging films: A critical review Food Packaging and Shelf Life 23: 100429
11. Zarrintaj P, Manouchehri S, Ahmadi Z, Saeb MR, Urbanska AM, Kaplan DL, Mozafari M, (2018) Agarose-based biomaterials for tissue engineering applications Carbohyd Polym 187:66-84
12. Cumpstey I (2013) Chemical modification of polysaccharides. ISRN organic chemistry
13. Li S et al (2016) Molecular modification of polysaccharides and resulting bioactivities. Compr Rev Food Sci Food Saf 15(2):237–250
14. Zeller SG, Greisgraber GW, Gray GR (1991) Analysis of positions of substitution of O-carboxymethyl groups in partially O-carboxymethylated cellulose by the reductive-cleavage method. Carbohyd Res 211(1):41–45

15. Machová E, Kvapilová K, Kogan G, Šandula J (1999) Effect of ultrasonic treatment on the molecular weight of carboxymethylated chitin–glucan complex from Aspergillus niger. Ultrason Sonochem 5(4):169–172
16. Maciel J, Silva D, Paula HC, De Paula R (2005) Chitosan/carboxymethyl cashew gum polyelectrolyte complex: synthesis and thermal stability. Eur Polymer J 41(11):2726–2733
17. Assaad E, Wang YJ, Zhu XX, Mateescu MA (2011) Polyelectrolyte complex of carboxymethyl starch and chitosan as drug carrier for oral administration. Carbohyd Polym 84(4):1399–1407
18. Mejía EH, Contreras H, Delgado E, Quintana G (2019) Effect of experimental parameters on the formation of hydrogels by polyelectrolyte complexation of carboxymethylcellulose, carboxymethyl starch, and alginic acid with chitosan. Int J Chem Eng
19. Mizrahy S, Peer D (2012) Polysaccharides as building blocks for nanotherapeutics. Chem Soc Rev 41(7):2623–2640
20. Aghadavoudi F, Golestanian H, Tadi Beni Y (2017) Investigating the effects of resin crosslinking ratio on mechanical properties of epoxy-based nanocomposites using molecular dynamics. Polymer Compo 38:E433–E442, 2017
21. Hu W, Wang Z, Xiao Y, Zhang S, Wang J (2019) Advances in crosslinking strategies of biomedical hydrogels. Biomaterials Sci 7(3):843–855
22. Friedman M (2012) Protein crosslinking: biochemical and molecular aspects. Springer Science & Business Media
23. Hecht H, Srebnik S (2016) Structural characterization of sodium alginate and calcium alginate. Biomacromol 17(6):2160–2167
24. Lee KY, Mooney DJ (2012) Alginate: properties and biomedical applications. Prog Polym Sci 37(1):106–126
25. Casassa E, Sarquis A, Van Dyke C (1986) The gelation of polyvinyl alcohol with borax: A novel class participation experiment involving the preparation and properties of a "slime". J Chem Educ 63(1):57
26. Mavila S, Eivgi O, Berkovich I, Lemcoff NG (2015) Intramolecular cross-linking methodologies for the synthesis of polymer nanoparticles. Chem Rev 116(3):878–961
27. Farokhi M, Mottaghitalab F, Fatahi Y, Saeb MR, Zarrintaj P, Kubdu SC, Khademhosseini A (2019) Silk fibroin scaffolds for common cartilage injuries: possibilities for future clinical applications Eur. Polym. J 115: 251-267
28. Khodadadi Yazdi M, Vatanpour V, Taghizadeh A, Taghizadeh M, Ganjali MR, Tajammal Munir M, Habibzadeh S, Saeb MR, Ghaedi M (2020) Hydrogel membranes: A review Mat Sci Eng C-Mater 114:111023
29. Miwa Y, Kurachi J, Kohbara Y, Kutsumizu S (2018) Dynamic ionic crosslinks enable high strength and ultrastretchability in a single elastomer. Commu Chem 1(1):5
30. Seidi F, Khodadadi Yazdi M, Jouyandeh M, Dominic M, Nasirnezhad M, Bagheri B, Habibzadeh S, Zarrintaj P, Saeb MR, Mozafari M (2021) Chitosan-based blends for biomedical applications Int J Biol Macromol 183: 1818-1850
31. Mecke A, Lee D-K, Ramamoorthy A, Orr BG, Banaszak Holl MM (2005) Synthetic and natural polycationic polymer nanoparticles interact selectively with fluid-phase domains of DMPC lipid bilayers. Langmuir 21(19):8588–8590
32. Li Y-Y, Wang B, Ma M-G, Wang B (2018) Review of recent development on preparation, properties, and applications of cellulose-based functional materials. Int J Polym Sci
33. Akalin GO, Pulat M (2018) Preparation and characterization of nanoporous sodium carboxymethyl cellulose hydrogel beads. J Nanomat
34. Wang G, Miao F, Dong F, Li Q, Guo Z, Guo Z (2014) Synthesis of Carboxymethyl Polysaccharides and their moisture absorption and retention abilities
35. Mocanu G, Souguir Z, Picton L, Le Cerf D (2012) Multi-responsive carboxymethyl polysaccharide crosslinked hydrogels containing Jeffamine side-chains. Carbohyd Polym 89(2):578–585
36. Li J, Lu J, Li Y (2009) Carboxylmethylcellulose/bentonite composite gels: water sorption behavior and controlled release of herbicide. J Appl Polym Sci 112(1):261–268
37. Butun S, Ince FG, Erdugan H, Sahiner N (2011) One-step fabrication of biocompatible carboxymethyl cellulose polymeric particles for drug delivery systems. Carbohyd Polym 86(2):636–643

38. Kadry G (2019) Comparison between gelatin/carboxymethyl cellulose and gelatin/carboxymethyl nanocellulose in tramadol drug loaded capsule. Heliyon 5(9):
39. El-Hag Ali A, Abd El-Rehim HA, Kamal H, Hegazy SA (2008) Synthesis of carboxymethyl cellulose based drug carrier hydrogel using ionizing radiation for possible use as site specific delivery system. J Macromol Science®, Part A: Pure Appl Chem 45(8):628–634
40. Hosny EA, Al-Helw AA-RM (1998) Effect of coating of aluminum carboxymethylcellulose beads on the release and bioavailability of diclofenac sodium. Pharm Acta Helv 72(5):255–261
41. Bulut E (2020) Chitosan coated-and uncoated-microspheres of sodium carboxymethyl cellulose/polyvinyl alcohol crosslinked with ferric ion: flurbiprofen loading and in vitro drug release study. J Macromol Sci, Part A 57(1):72–82
42. Bulut E, Şanli O (2014) Optimization of release conditions of Alzheimer's drug donepezil hydrochloride from sodium alginate/sodium carboxymethyl cellulose blend microspheres. J Macromol Sci, Part 53(5):902–917
43. Kim MS, Park SJ, Gu BK, Kim C-H (2012) Ionically crosslinked alginate–carboxymethyl cellulose beads for the delivery of protein therapeutics. Appl Surf Sci 262:28–33
44. Samrot AV, Jahnavi T, Padmanaban S, Philip S-A, Burman U, Rabel AM (2016) Chelators influenced synthesis of chitosan–carboxymethyl cellulose microparticles for controlled drug delivery. Appl Nanosci 6(8):1219–1231
45. Kahya N, Gölcü A, Erim FB (2019) Barium ion cross-linked alginate-carboxymethyl cellulose composites for controlled release of anticancer drug methotrexate. J Drug Deliv Sci Technol 54:
46. Dey P, Maiti S, Sa B (2013) Gastrointestinal delivery of glipizide from carboxymethyl locust bean gum–Al^{3+}–alginate hydrogel network: In vitro and in vivo performance. J Appl Polym Sci 128(3):2063–2072
47. Abdelgawad A, Hudson S (2019) Chitosan nanoparticles: Polyphosphates cross-linking and protein delivery properties. Int J Bio Macromolecules
48. Quiñones JP, Peniche H, Peniche C (2018) Chitosan based self-assembled nanoparticles in drug delivery. Polymers 10(3):235
49. Kalliola S et al (2017) The pH sensitive properties of carboxymethyl chitosan nanoparticles cross-linked with calcium ions. Colloids Surf, B 153:229–236
50. Lu Y et al (2017) Multifunctional copper-containing carboxymethyl chitosan/alginate scaffolds for eradicating clinical bacterial infection and promoting bone formation. ACS Appl Mater Interfaces 10(1):127–138
51. Mi F-L, Chen C-T, Tseng Y-C, Kuan C-Y, Shyu S-S (1997) Iron (III)-carboxymethylchitin microsphere for the pH-sensitive release of 6-mercaptopurine. J Controlled Release 44(1):19–32
52. Haroon M et al (2016) Chemical modification of starch and its application as an adsorbent material. Rsc Adv 6(82):78264–78285
53. Saikia C, Hussain A, Ramteke A, Sharma HK, Deb P, Maji TK (2015) Carboxymethyl starch-coated iron oxide magnetic nanoparticles: a potential drug delivery system for isoniazid. Iran Polym J 24(10):815–828
54. Ispas-Szabo P, De Koninck P, Calinescu C, Mateescu MA (2017) Carboxymethyl starch excipients for drug chronodelivery. AAPS PharmSciTech 18(5):1673–1682
55. Mohapatra S et al (2019) Doxorubicin loaded carboxymethyl Assam bora rice starch coated superparamagnetic iron oxide nanoparticles as potential antitumor cargo. Heliyon 5(6):
56. Zhang B, Tao H, Wei B, Jin Z, Xu X, Tian Y (2014) Characterization of different substituted carboxymethyl starch microgels and their interactions with lysozyme. PLoS One 9(12):
57. Zhang B, Wei B, Hu X, Jin Z, Xu X, Tian Y (2015) Preparation and characterization of carboxymethyl starch microgel with different crosslinking densities. Carbohyd Polym 124:245–253
58. Saboktakin MR, Tabatabaie RM, Maharramov A, Ramazanov MA (2011) Synthesis and in vitro evaluation of carboxymethyl starch–chitosan nanoparticles as drug delivery system to the colon. Int J Biol Macromol 48(3):381–385

59. Quadrado RF, Fajardo AR (2018) Microparticles based on carboxymethyl starch/chitosan polyelectrolyte complex as vehicles for drug delivery systems. Arabian J Chem
60. Ning S et al (2011) Carboxymethyl dextran-coated liposomes: Toward a robust drug delivery platform. Soft Matter 7(19):9394–9401
61. Thambi T et al (2014) Bioreducible carboxymethyl dextran nanoparticles for tumor-targeted drug delivery. Adv Healthcare Mat 3(11):1829–1838
62. Goodarzi N, Varshochian R, Kamalinia G, Atyabi F, Dinarvand R (2013) A review of polysaccharide cytotoxic drug conjugates for cancer therapy. Carbohyd Polym 92(2):1280–1293
63. Lin Y, Radzi R, Morimoto M, Saimoto H, Okamoto Y, Minami S (2012) Characterization of chitosan-carboxymethyl dextran nanoparticles as a drug carrier and as a stimulator of mouse splenocytes. J Biomater Sci Polym Ed 23(11):1401–1420
64. Collins MN, Birkinshaw C (2013) Hyaluronic acid based scaffolds for tissue engineering—A review. Carbohyd Polym 92(2):1262–1279
65. Tiwari S, Bahadur P (2019) Modified hyaluronic acid based materials for biomedical applications. Int J Biol Macromol 121:556–571
66. Huang G, Huang H (2018) Application of hyaluronic acid as carriers in drug delivery. Drug Deliv 25(1):766–772
67. Kim K, Choi H, Choi ES, Park M-H, Ryu J-H (2019) Hyaluronic acid-coated nanomedicine for targeted cancer therapy. Pharmaceutics 11(7):301
68. Woodman JL et al (2015) Carboxymethyl hyaluronan-stabilized nanoparticles for anticancer drug delivery. Int J Cell Biology
69. Jun W, Lin L, Yurong C, Juming Y (2013) Recent advances of calcium phosphate nanoparticles for controlled drug delivery. Mini Rev Med Chem 13(10):1501–1507
70. Thimma RT, Tammishetti S (2001) Barium chloride crosslinked carboxymethyl guar gum beads for gastrointestinal drug delivery. J Appl Polym Sci 82(12):3084–3090
71. Maity S, Sa B (2014) Ca-carboxymethyl xanthan gum mini-matrices: swelling, erosion and their impact on drug release mechanism. Int J Biol Macromol 68:78–85

Chapter 6
Polyelectrolyte Complex-Based Ionically Gelled Biopolymeric Systems for Sustained Drug Release

V. Ponnusami

1 Introduction

The therapeutic agents such as proteins, peptides, or even cells are generally not stable at extreme conditions prevailing in the gastrointestinal tract. An ideal drug delivery system should protect the drugs from unfriendly environments and deliver them at the specific site, at the desired rate and in appropriate dosage forms [1–3]. For example, oral drug delivery systems must be designed in such a way that the drug carrier has good resistance against gastrointestinal enzymes and pH gradients (i.e., from 1 to 3 in the stomach to 6 to 7 in the intestine) [4]. The choice of the drug carrier to meet these requirements is critical in the design of drug delivery systems. Various polymeric substances can be used as a drug career. Use of natural polysaccharides is gaining attraction recently, as they are biodegradable, biocompatible, and less toxic. Apart from this, natural polysaccharides are amenable to simple modifications through which their properties can be fine-tuned to meet the specific requirements for a given therapeutic application; they can absorb drug molecules more efficiently and release them at a controlled rate. Polysaccharides from various natural sources such as algal sources (e.g., alginate), plant source (e.g., pectin, guar gum), microbial origin (e.g., cellulose, dextran, pullulan, levan, xanthan gum, gellan gum, etc.), and animal orgin (e.g., chitin, chitosan) are employed in drug delivery applications includes [2].

Recently, ionic gelation is gaining the attention of researchers as a drug delivery system formulation procedure. The chemical crosslinkers and organic solvents used in conventional methods of producing drug delivery systems may result in damage to the active molecules like DNA, proteins, and peptides and/or reduce their therapeutic efficiency. The presence of even traces of solvents in the drug dosage form

V. Ponnusami (✉)
Biomass Conversion and Bioproducts Laboratory, Center for Bioenergy,
School of Chemical and Biotechnology, SASTRA Deemed University, Thanjavur 613401, India
e-mail: vponnu@chem.sastra.edu

A. K. Nayak et al. (eds.), *Ionically Gelled Biopolysaccharide Based Systems in Drug Delivery*, Gels Horizons: From Science to Smart Materials,
https://doi.org/10.1007/978-981-16-2271-7_6

is objectionable in many drug delivery applications [2]. Ionic gelation is a simple method which employs only mild conditions and does not require the use of organic solvents or ultrasonication. Thus, it is considered as a safe method of synthesizing polysaccharide nanoparticles for drug delivery applications. This chapter provides an overview of various aspects of polyelectrolyte complex-based drug delivery systems.

2 Ionic Gelation and Polyelectrolyte Complexation

Most of the natural polysaccharides are polyionic in nature. Biopolysaccharides such as chitosan, dextran, starch, and cellulose are examples of polycations while alginate, carrageenan, carboxymethyl cellulose, gelatin, gellan gum, hyaluronic acid, pectin, xanthan gum, and heparin are examples of polyanions [5]. Thus, they have the ability to form a gel in the presence of suitable counterions through ionic crosslinking. Counterions used in ionic gelation, based on their molecular weight, can be classified into two categories, namely (i) high molecular weight counterions and (ii) low molecular weight counterions. Tripolyphosphate (TPP) is the most commonly used low molecular weight crosslinking agents in ionic gelation [6]. It is multivalent and non-toxic. Polyanions such as carboxymethyl cellulose, carboxymethylated starch, and alginate can be ionically crosslinked with low molecular weight polycations like calcium chloride. Similarly, polycations (e.g., chitosan) are crosslinked with low molecular weight polyanions like TPP. Typically, this type of crosslinking is termed as ionic gelation. Other than TPP, pentasodium tripolyphosphate [7], sodium polyphosphate [8], sodium sulfate, molybdate were also used as gelation agents by researchers [1]. Nanoparticles obtained from few polysaccharides like sodium carboxymethyl xanthan gum and pectin are not stable at physiological conditions leading to a rapid release of the drugs and hence result in poor therapeutic efficacy. In order to overcome this drawback, polyelectrolyte complexation between various biopolymers had been tried and shown to be more efficient in many cases.

While neutral polymers occupy less space through random coil conformation in aqueous solutions, polyelectrolytes stretch out more in aqueous solution due to the repulsion between like charges. When two oppositely charged polymer (for e.g., chitosan [polycation] and carboxymethyl cellulose [polyanion]) solutions are mixed properly, ionic crosslinking between oppositely charged groups are formed due to one or more ionic interactions such as electrostatic and dipole-dipole interaction, hydrogen bonding, and hydrophobic interactions. This process is termed as polyelectrolyte complexation [9–11]. Polyelectrolyte complexation between oppositely charged polymers results in a stable complex [12] and exhibits better mechanical and drug release properties. Also, polyelectrolyte complexes are less affected by pH variation and hence more suitable for sustained delivery of drugs at physiological conditions. Polyelectrolyte complexation can be used for encapsulation, transportation, and delivery of numerous bioactive agents [11].

When two oppositely charged polyionic polymers are mixed, the aggregation of particles and the formation of insoluble cooperative complexes should occur at a thermodynamically stable condition. The electrostatic interactions between oppositely charged functional groups, Van der Waals force, and hydrogen bonding are responsible for the structural transformation taking place during these reactions. Thus, thermodynamic stability is achieved at some critical ratios of charges on the interacting polymers [13]. Factors influencing the formation of polyelectrolyte complex can be broadly put under three categories: (i) structural parameters (charge stoichiometry, molecular confirmation of polymers, charge density distribution of ionic groups of the polymers, molecular weight of the polymers), (ii) media parameters (solution pH, concentrations of the polymers, ionic strength, and salt concentration), and (iii) preparation parameters (solution temperature, order of mixing, mixing ratio, reaction time [5]. All these factors affect the thermodynamic stability of the system. Depending on these conditions, either the solvation of the polymers takes place or phase separation occurs due to polyelectrolyte complexation [14]. During polyelectrolyte complexation, the accessibility of the opposite charge could be different depending on the position of the charged groups on the polymer chains and are affected by steric hindrance. According to the mode of preparation, different types of polyelectrolyte complexes structure like "ladder" or "scrambled eggs" may be obtained during polyelectrolyte complexation [15]. Typically, when the charge ratio of the polymers is equal the total charge is zero, it leads to aggregation and macroscopic phase separation.

On the other hand, soluble polyelectrolyte complexes are obtained with an excess of either anionic or cationic charges leading to the formation of stable and transparent solutions [16]. Thus, smaller particles are obtained when one of the polymers is used in excess [10]. The surface morphology and functional properties of reaction products (fibers, films, gels) are thus largely determined by the thermodynamics of polyelectrolyte complex formation. Some common polyelectrolyte complexes obtained from natural polysaccharides used in drug delivery applications are listed in Table 1.

In some cases, the ionic bridge formed between two oppositely charged polymers through ionic crosslinking using low molecular weight polyions can be used as a drug delivery system [12, 36]. It is possible to modulate the electrostatic interaction in a polyelectrolyte solution by adding suitable salts at a required concentration which changes the ionic strength of the polymers. An increase in ionic strength of the solution by the addition of low molecular weight counterions leads to the screening of opposite charges of the macromolecules. This can weaken the complexation between polyions by weakening the electrostatic interactions between functional groups of the biopolymers. The degree of ionization of weak polyelectrolytes can also be controlled by adjusting the solution pH during PEC formation [12]. A schematic representation of gelation reaction is shown in Fig. 1a, b. When an aqueous solution of water-soluble polysaccharides is added dropwise to a solution containing counterions under constant stirring, crosslinking between drug-loaded polysaccharide and its counterions occurs under mild conditions [37].

Characterization of the resulting particles/gel is generally performed using FTIR, SEM, zeta potential, and DSC to study the interaction between the polymers,

Table 1 Polyelectrolyte complexes of natural polysaccharides in drug delivery applications

Polycation	Polyanion	Bioactive component career/application	Formulation	References
Chitosan	Hyaluronic acid	Anti-ovalbumine (OVA) Immunoglobulin A (IgA)	NP	[17]
	Carboxy methyl cellulose	Vancomycin	Microparticles	[14]
	Pectin, gum Arabic	–	Membrane	[18]
	Dextran sulfate/heparin		Macrohydrogel/colloidal gel	[19]
	Dextran sulfate		Core-shell (excess component forms outer shell)	[20]
	Pectin	Vancomycin	Gel	[21]
	Carboxymethyl dextran	Gene	NP	[22]
	Sodium alginate	Tissue engineering and drug delivery	Fibrous structure at low pH; nano-colloids at high pH	[23]
	Hyaluronic acid	Drug delivery	NP	[24]
	Gelatin	–	Aqueous phase complex	[25]
	Alginate	Insulin	NP	[26]
	Sodium alginate		Gel	[13]
	Pectin	Tacrine hydrochloride	Microparticles	[27]
	Dextran sulfate	Drug delivery	Colloidal PEC	[28]
	Xanthan gum	Tissue engineering	Hydrogel	[29]
	Gum arabic	Curcumin	NP	[30]
	Hyaluronic acid	–	Colloidal suspensions/gel coacervates	[31]
	Alginate	–	Polyelectrolyte aggregates	[32]
	Hyaluronic acid	–	Polyelectrolyte complex film	[33]
	Carboxymethyl cellulose	Drug delivery	Microparticles	[34]
Chitosan hydrochloride	Carboxymethyl starch	Curcumin	Nanogels	[35]
Cationized gelatin	Sodium alginate	Curcumin	NP	[11]

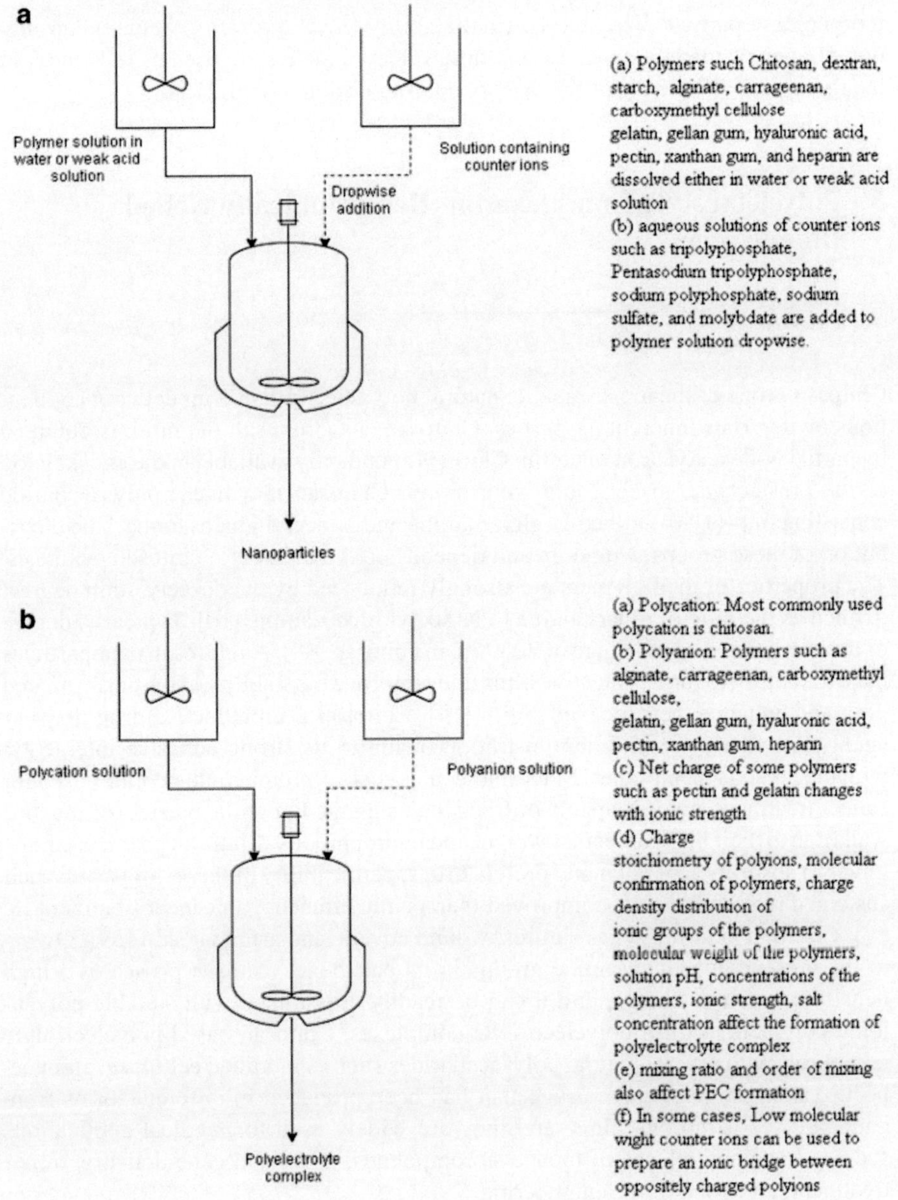

Fig. 1 **a** Schematic diagram for ionic gelation and **b** schematic diagram for polyelectrolyte complexation

morphology, particle size, and thermal stability. Swelling ratio, yield, encapsulation efficiency, loading capacity, and drug release kinetics are used as indicators of the therapeutic efficiency of the developed drug design system [13].

3 Polyelectrolyte Complexation-Based Ionically Gelled Biopolymers

3.1 Chitosan

Chitosan is one of the most versatile natural polymers used in drug delivery applications owing to its inherent properties. Chitosan, a sodium salt of chitin, is obtained by partial N-deacetylation of chitin. Chitin is abundantly available in the exoskeleton of the crustaceans, insects, and some fungi. Chitosan is a linear polysaccharide consisting of β-(1→4) linked D-glucosamine and N-acetyl glucosamine. The distribution of these groups is random and depends on the method of chitosan synthesis. The properties of the polymers are strongly influenced by the deacetylation degree, defined as the ratio of glucosamine to N-acetyl glucosamine [10]. Typically, degree of deacetylation is in the range of 70–95% in chitosan [38]. As chitosan nanoparticles have the ability to form protective films, they improve the stability of the encapsulated drug and enhance their bioavailability [14]. Chitosan is an attractive drug delivery agent particularly for oral administration owing to its strong adhesive interaction with the mucous membranes. Macromolecules such as proteins and peptides used for cancer treatment are not capable of freely crossing the lipophilic barriers of mucous walls due to their high molecular weight and hydrophilicity. Chitosan promotes transmucosal absorption of polar and protein drugs, permeability of these drugs, aids their sustained release to achieve improved therapeutic efficiency in cancer treatment [8, 39]. Chitosan is also used as sutures, wound covers, and artificial skin [39]. Owing to the presence of glucosamine groups in its backbone, chitosan possesses a high density of positive charge, and it can be readily crosslinked with suitable polyanionic polymers to form polyelectrolyte complexes. Chitosan-based polyelectrolyte complexes with several natural polysaccharides such as modified cellulose, alginate, pectin, xanthan gum, and carrageenan had been produced in different forms from nanoparticles to fibrous films, and they are widely used for medical applications such as sustained release of bioactive compounds, DNA and gene delivery, tumor treatment, and bone tissue engineering.

Concentration of chitosan, concentration of crosslinker, and chitosan to crosslinker ratio significantly affect physicochemical characteristics of chitosan nanoparticles produced. Also, with an increase in drug concentration, particle size increased and entrapment efficiency decreased [7]. Chitosan and its derivatives diethyl methyl chitosan (DEMC) and trimethyl chitosan (TMC) were employed for insulin delivery by Sadeghi et al. [39]. Chitosan was dissolved in 0.25% acetic acid, and the solution pH was adjusted to 5.0. At this solution pH, most of the amine groups in chitosan were protonated. Insulin solution with 1 g/L concentration prepared in 0.01 N HCl solution is then added to the polymer in 1:1 ratio under constant stirring at room temperature to form polyelectrolyte complex loaded with insulin. Polyelectrolyte complex of chitosan derivatives was obtained by following the same procedure with an aqueous solution of DEMC or TMC instead of acidic chitosan solution. The authors compared the characteristics and performance of nanoparticles synthesized by both polyelectrolyte complexation and ionic gelation. Among all the six nanoparticles used, chitosan nanoparticles synthesized by polyelectrolyte complexation method had shown maximum insulin loading. Insulin loading depends on the ionic interaction between the drug and the carrier. In polyelectrolyte complexation method, negatively charged insulin directly interacts with positive charges of the polymer. As the net positive charge on unmodified chitosan was higher than that on chitosan derivatives at pH 5 the ionic interaction was stronger in unmodified chitosan, and therefore, the drug loading was also higher in unmodified chitosan than its derivatives. Also, the poor interaction between insulin with quaternized derivatives was attributed to the fact that the methyl and ethyl groups shield the positively charged nitrogen ions of the polymer. On the other hand, in ionic gelation, insulin does not interact with polymer directly. As insulin binds with chitosan through the low molecular weight crosslinker TPP, the drug tends to diffuse into the external medium more easily leading to less drug loading than that obtained in polyelectrolyte complexation method. The zeta potential, which indicates the available surface charge of the nanoparticle, was high for nanoparticle obtained by PEC compared to that obtained by ionic gelation. For the drug delivery application, the net positive charge of the nanoparticle carrier is preferred to enhance the interaction with cellular membrane components and permeation of the drug across intestinal epithelium [39]. However, TMC particles prepared by PEC had the highest zeta potential and antimicrobial inhibition among the six careers studied by the authors [39]. It has also been reported that compared to chitosan which required weak acidic medium during nanoparticle synthesis, water-soluble chitosan derivatives exhibit better preservation of bioactivity of drug molecules. [40].

When chitosan is partially sulfated, under appropriate conditions, polyelectrolyte complexes are formed due to the electrostatic interactions between the positively charges non-sulfated amines and the negatively charged N-sulfated amines. Macromolecular drugs can be encapsulated with this polyelectrolyte complexes to design a controlled drug delivery system [9]

3.2 Alginate

Alginate is a natural polysaccharide present in the cell walls of brown algae as sodium, potassium, or calcium salts of alginic acid. It is a linear polysaccharide containing two uronic acids, namely α-L-glucuronic acid, and β-D-mannuronic acids [1]. Alginate has a net negative charge due to the presence of carboxyl groups of uronic acids. It is ionically crosslinked with low molecular weight cations like Ca^{2+}, Ba^{2+}, Zn^{2+}, or high molecular weight polycation like chitosan to form efficient drug delivery systems. Since the binding between alginate and low molecular weight cations is weak/loose, the hydrogels produced by ionic gelation are unstable. On the other hand, the binding is stable in polyelectrolyte complexes formed with polycation and polyanion polymers because of strong electrostatic interaction between amino groups of chitosan and carboxyl groups of alginate [10, 41]. Alginate polyelectrolyte complexes obtained with polycations such as chitosan are capable of protecting these usually unstable molecules, maintaining their bioactivity and enhancing their absorption and permeation through various mucosal surfaces and biological membranes when administered through oral routes [10].

Alginate–chitosan polyelectrolyte complexes are recognized as promising drug delivery systems for macromolecular drugs such as insulin [42] proteins, peptides, and nucleotides. To produce alginate–chitosan polyelectrolyte complex typically, alginate is first added to a solution containing calcium ion to initiate gelation. Subsequent addition of chitosan to the mixture leads to the formation of alginate–chitosan nanoparticle through PEC [43]. Typically, the pKa value of alginate is in the range of 3.3–3.7, and the pKa of chitosan is about 6.5. To produce PEC, the solution pH must be between pKa values of chitosan and alginate [32]. Similarly, a low molecular weight cation TPP has also been used to reinforce the alginate–chitosan polyelectrolyte complexes. Introducing TPP in alginate–chitosan polyelectrolyte complexes altered the supermolecular structure and swelling behavior favorably [44].

Polyelectrolyte complexation of alginate with another polyanion dextran sulfate in the presence of chitosan and calcium cations also found to improve insulin loading capacity, entrapment efficiency, and insulin retention [4]. Alginate-based polyelectrolyte complexes formed with polycationic polymers such as pectin, gelatin [11], and ethyl cellulose had been employed for different drug delivery applications.

Henao et al. [36] synthesized chitosan-based hydrogel by: (1) ionic gelation with $CaCl_2$ as crosslinker, (2) polyelectrolyte complexation with polyanionic polymers carboxymethylcellulose, carboxymethyl starch, and alginic acid, and (3) a combination of these methods using both $CaCl_2$ and carboxymethylcellulose as crosslinkers [36]. The authors had reported that the morphology of the chitosan-based hydrogel formed was influenced by the ionic crosslinking process employed for the synthesis. In ionic gelation, when $CaCl_2$ was used as the crosslinker, at low concentrations of $CaCl_2$,, a transparent hydrogel was obtained, and at higher $CaCl_2$ concentration, a cloudy gel with conical and spiral-shaped particles was obtained. Meanwhile, when polyanionic crosslinking agents were used, a sponge-like membrane was produced by polyelectrolyte complexation. In simultaneous PEC and ionic gelation method, cloudy and viscous suspension of elongated particles was obtained. In the case of simultaneous ionic gelation and polyelectrolyte complexation, $CaCl_2$ enables better interaction between cationic and anionic polymers by acting as a bridge between these two polymers [36].

3.3 Carrageenan

Carrageenan is a sulfated linear polysaccharide commonly found in red algae. Carrageenan is composed of galactose and anhydrous galactose units with glycosidic links [10]. Depending on the source and the extraction method, four types (κ-, κ/β-, λ, and χ) of carrageenan with different degrees of sulfonation had been reported. Carrageenan has a negative charge, and it shows a high affinity to polycationic polysaccharides like chitosan and forms polyelectrolyte complexes. As both carrageenan and chitosan have wide spectrum antibiotic activity, carrageenan–chitosan polyelectrolyte complexes are suitable for several drug delivery applications. Carrageenan–chitosan polyelectrolyte complexes are produced in different forms such as films, microcapsules, microspheres, and gels [45]. Carrageenan–chitosan PEC had been shown to have gastroprotective activity. Volod et al. [45] synthesized PEC of carrageenan obtained from *Chondrus armatus* by mixing it with a solution of chitosan at ratios of 1:10 and 10: 1 w CG/w Chitosan. The PEC with 1:10 CG to chitosan ratio had exhibited better gastroprotective activity. This was attributed to the favorable supramolecular structure of the PEC obtained at 1:10 ratio [45]. In a follow-up study, Volod et al. [46] had synthesized polyelectrolyte complexes of κ-, κ/β-, λ, and χ–types CG with chitosan. The type of CG was found to affect the complex formation significantly. The binding affinity for chitosan was strong for χ-CG than other forms of CG, and a higher degree of sulfonation in χ-CG was identified as the reason for this observation [46].

3.4 Cellulose

Cellulose is the most abundant polymer in nature. It is a linear polysaccharide composed of β (1→4) linked glucose monomers [47]. Chitosan–carboxy methylcellulose gels had been recently used for site-specific drug delivery of different drugs. Since vancomycin, a glycopeptide antibiotic which is used for the treatment of infections caused by species like *Staphylococcus*, is not stable under conditions prevailing in the stomach, it is usually administered through intravenous systems [14]. To prevent side effects associated with this, carboxymethyl cellulose–chitosan polyelectrolyte complexes were used as a drug delivery system for oral administration vancomycin for colon-specific infections [14]. Polyelectrolyte complexes with different chitosan: carboxymethylcellulose ratios of 3:1, 1:1 and 1:3 were prepared, and PEC with 1:3 was found to show superior performance. Drug release was negligible at pH 2.0, but at a pH of 7.4, significant drug release was obtained, thus making it a promising oral drug delivery system [14].

3.5 Curdlan

Curdlan is a water-insoluble linear polysaccharide consisting of 1→3 β-D-glucan. Sulfate modification of curdlan makes it water soluble and suitable for drug delivery applications. Curdlan sulfate is recognized as an effective anti-HIV agent, owing to its antiviral activity, and it is in phase II clinical trials. Negatively charged sulfate groups of curdlan sulfate form polyelectrolyte complexes with protonated amine groups of chitosan. Curdlan–chitosan polyelectrolyte complexes were used for delivery of zidovudine (antiviral) [48] and 5-fluorouracil (anti-tumor) [49].

3.6 Dextran

Dextran is a branched polysaccharide made glucose monomers. The main chain has α (1→6) glycosidic linkages between glucose units. The branch chains are α (1→3) linked to the main chain. Krauland and Alonso [50] employed chitosan nanoparticles synthesized by ionic gelation technique using TPP and carboxymethyl-β-cyclodextrin as crosslinkers. The size and zeta potential of the produced nanoparticles were in the range of 231–383 nm and +20.6 to +39.7 mV. The stability of the particles was evaluated at simulated intestinal conditions of pH 6.8 and temperature 37 C, and the particles were found to be stable for at least 4 h. Heparin and insulin loading efficiency on these nanoparticles were in the range of 69.3–70.6% and 85.5–93.3%, respectively. Loading of these molecules resulted in an increase in particle size up to 613 nm. However, zeta potential did not change significantly. The release rate was rapid for insulin, and about 84–97% of insulin was released within 15 min. But, the release rate was very slow for heparin. Even after 8 h, only 8.3–9.1% of heparin was released [50].

It has also been reported that the addition of salts like NaCl is useful to control the formation of chitosan–dextran sulfate/heparin polyelectrolyte complexes. The nature of the product formed depends on the stoichiometric ratio of polycation (n^+) and polyanion (n^-). If this ratio $n^+/n^- = 1$, flocculation occurs. For ratio less than or greater than 1, phase separation occurs and colloids are formed [19, 51]. At a salt concentration of 2 mol L^{-1} NaCl, complete control over these ionic interactions was achieved. With an excess of polyanions, electrostatic crosslinking was faster which led to the formation of macrohydogels. On the other hand, excess of polycations resulted in colloidal gels due to conformational adaptation and chain disentanglement. At a salt concentration of less than 1 mol L^{-1}, macrohydrogels were obtained spontaneously irrespective of polyion concentration [19].

3.7 Gelatin

Gelatin exists in two forms, namely Type A and Type B. The isoelectric point of gelatin is in the range of 7–9 and 4.7–5.2 for type A gelatin and type B gelatin, respectively. Thus, depending on the medium, pH gelatin is either positively charged or negatively charged and forms polyelectrolyte complexes with respective counter polyions. For example, cationically modified gelatin is crosslinked with anionic polysaccharides dextran sulfate and chondroitin sulfate, and the resulting polyelectrolyte complexes were used in ocular gene delivery [11]. Gelatin–alginate polyelectrolyte complexes had been used for the delivery of drugs ciprofloxacin and curcumin [11]. On the other hand, anionic gelatin forms polyelectrolyte complexes with cationic polymers such as chitosan [25].

3.8 Hyaluronic Acid

Hyaluronic acid is another natural biopolymer found in connective, epithelial, and neural tissues of various organisms. In human body, HA is widely present in skin, cartilage, synovia, and vitreous tumor [31]. It consists of β-1,4-D-glucuronic acid—β-1,3—N-acetyl—D-glucosamine repeating units. The molecular weight of HA ranges from 100 kDa to 8000 kDa. Above its pKa value (2.49), hyaluronic acid has an anionic charge (due to the carboxyl groups) and can form polyelectrolyte complexes with cationic polysaccharides like chitosan [31]. The rate of chitosan–hyaluronic acid complexation needs to be controlled to make materials suitable for biomedical applications. Lalevee et al. [31] employed charge screening successfully to achieve this control [31]. Colloidal stability, low toxicity, ability to protect the drug molecules from pH variations and enzymatic degradation, tunable size make hyaluronic acid-based polyelectrolyte complexes a promising drug delivery system. The chitosan–HA polyelectrolyte complexes had been used as a drug delivery system for oral and nasal drug delivery systems. It has been reported that the performance of the chitosan–HA polyelectrolyte complexes is superior compared to many other polyelectrolyte complexes for drug delivery application due to high mucoadhesive property of HA and enhanced penetration effect of chitosan [10].

Wu et al. [52] designed a novel colloidal drug delivery system consisting of a ternary PEC chitosan–heparin sodium salt–hyaluronic acid. First, a polyanionic heparin sodium salt to chitosan solution to form a PEC. Further addition of hyaluronic acid to the resultant product to form a PEC consisting of chitosan, heparin sodium salt, and hyaluronic acid. Depending on the ratio of chitosan–heparin to hyaluronic acid, the resulting PEC was either cationic or anionic. The authors also had shown similar results by replacing heparin sodium salt with dextran sulfate [52]. In earlier work, Wu et al. [24] had obtained polyelectrolyte complexes of hyaluronan with chitosan and had shown that the addition of Zn^{2+} improved colloidal stability of the polyelectrolyte complexes [24].

Lalevee et al. [31] designed a PEC between polycation chitosan (CS) and polyanion hyaluronic acid (HA) which is stable at physiological salt concentration and pH. Nanoparticles were obtained by charge neutralization between these oppositely charged polyions. As mentioned earlier, the degree of acetylation, molar mass of CS, the HA molar mass, the charge mixing ratio, and the polymer concentrations affect the supramolecular and physiochemical properties of the polyelectrolyte complexes produced. Non-stoichiometric colloidal polyelectrolyte complexes obtained in water or physiological buffer (PBS) at pH 7.4 were reported to be stable for more than a month. The average size of CS-HA polyelectrolyte complexes loaded with antibody anti-ovalbumin (OVA) immunoglobulin A (IgA) was found to be in the range of 425–665 nm, and zeta potential was positive. The CS-HA/IgA system had shown fast drug release kinetics with high drug loading capacities. [17].

3.9 Pectin

Natural polysaccharide pectin is obtained from the cell walls of all higher plants. Pectin refers to a family of oligosaccharides and polysaccharides. The chemical structure of pectin is diverse and depends on its source. Typically, it is rich in α-(1→4) D-galactosyluronic acid residues with acetyl and methyl esterified carboxyl groups. The net surface charge and degree of acetylation and methylation determine its functional properties. Though pectin has a good tendency to form gels, the application of pectin hydrogels is limited in drug delivery application as it swells a lot under physiological conditions. Pectin is stable in the gastrointestinal tract but degrades well in the colon. Because of this reason, pectin is suitable for colon targeted drug delivery systems.

Pectin when crosslinked with divalent Ca^{2+} ions forms an insoluble complex with "Egg box" structure. Calcium pectinate gel obtained by ionotropic gelation had been used for the delivery of therapeutic agents such as indomethacin, sulphamethoxazole, and metronidazole [53]. Owing to their opposite surface charges, pectin and chitosan have a good tendency to form PEC through ionic crosslinking. The pKa value of carboxyl groups of pectin is in the range of 6–8. For the amino groups in chitosan, the pKa value is around 6.5. The carboxyl groups of pectin tend to interact with protonated amine groups of chitosan to form hydrogel. da Costa et al. [54] produced pectin–chitosan PEC in the presence of montmorillonite clay in an attempt to reduce the production cost. The resulting PEC exhibited improved water absorption capacity, swelling ability, mechanical strength, and thermal stability [54].

When a mixture of pectin and sodium carboxymethyl xanthan was brought in contact with Al^{3+} ions, the cations readily formed crosslinks between the two biopolymers [53]. The size of the gel beads formed was a function of drug loading. The size of the gel bead increased with an increase in drug loading. This was attributed to the increase in interfacial viscosity associated with an increase in drug loading. The concentration of $AlCl_3$ in the solution affects the crosslinking density positively [53].

3.10 Starch

Polyelectrolyte complex produced from anionic carboxymethyl starch (CMS) using cationic chitosan has been found to be another effective drug delivery system. Compared to TPP, chitosan had a better ionic interaction with CMS. CMS–chitosan polyelectrolyte complexes exhibited high drug loading and pH-dependent drug release kinetics. Thus, CMS–chitosan polyelectrolyte complexes had been recognized as a promising drug delivery system for sustained drug release [55].

3.11 Xanthan Gum

Xanthan gum is a microbial exopolysaccharide produced by *Xanthomonas campesteris*. Xanthan gum has anionic charge due to the pyruvic and acetyl residues attached to its side chains. Like other polyelectrolyte complexes discussed above, xanthan gum–chitosan polyelectrolyte complexes also provide stability to the drug molecules under gastrointestinal conditions and improve bioavailability and sustained release of the drug.

Xanthan gum–chitosan polyelectrolyte complexes are formed due to the ionic interaction between amine groups of chitosan and carboxyl groups of xanthan gum. These gels are used for sustained release of drugs, particularly for those administered orally [56]. These gels are non-toxic, have high resistance toward enzymes present in digestive systems, and display pH-sensitive swelling characteristics. The properties of the gels formed can be easily tuned by controlling the gelation conditions such as pH, concentration of polymers and their mass ratio, complexation time and adjusting the molecular properties of the polymers such as the degree of acetylation of chitosan, pyruvic acid content of xanthan, and molecular weight [56]. Kem et al. synthesized chitosan–xanthan gum gel by polyelectrolyte precipitation and employed for the delivery of a broadspectrum antibiotic and antiseptic drug, chlorhexidine. By changing the ratio of chitosan to xanthan gum, pH and concentration of chitosan gels with different morphological and swelling properties were obtained. Antibacterial activity of the resulting gels was demonstrated through *in vitro* tests [57]. Madhusudana et al. (2018) had recently demonstrated that by incorporation of iron oxide nanoparticles, xanthan gum–chitosan PEC can be used effectively for magnetically stimulated drug delivery systems [29]. Xanthan gum–chitosan formed with different cation to anion ratios was employed for the delivery of high dosage drug paracetamol. Modified xanthan gum (e.g., sodium carboxymethyl xanthan gum) forms a gel with cations (e.g., Al^{3+}) through ionic gelation. Sodium carboxymethyl xanthan has been used to encapsulate ibuprofen, diltiazem hydrochloride, bovine serum albumin for sustained drug delivery [53].

References

1. Racovita S, Vasiliu S, Popa M, Luca C (2009) Polysaccharides based on micro- and nanoparticles obtained by ionic gelation and their applications as drug. Rev Roum Chim 54(9):709–718
2. Smelcerovic A, Knezevic-jugovic Z, Petronijevic Z (2008) Microbial Polysaccharides and their derivatives as current and prospective pharmaceuticals. Curr Pharm Des 14:1–28
3. Usmiati S, Richana N, Mangunwidjaja D, Noor E, Prangdimurti E (2014) The Using of Ionic Gelation Method Based on Polysaccharides for Encapsulating the Macromolecules—A Review 2. Encapsulation for Protecting the Bioactive Compounds. Int Conf Food Secur Nutr 67:79–84
4. Martins S, Sarmento B, Souto EB, Ferreira DC (2007) Insulin-loaded alginate microspheres for oral delivery—Effect of polysaccharide reinforcement on physicochemical properties and release profile. Carbohydr Polym 69:725–731
5. Kulkarni AD et al (2016) Polyelectrolyte complexes: mechanisms, critical experimental aspects, and applications. Artif Cells, Nanomedicine, Biotechnol 44(7):1615–1625
6. Jain A, Thakur K, Sharma G, Kush P, Jain UK (2016) Fabrication, characterization and cytotoxicity studies of ionically cross-linked docetaxel loaded chitosan nanoparticles. Carbohydr Polym 137:65–74
7. Anal AK, Stevens WF, Remu C (2006) Ionotropic cross-linked chitosan microspheres for controlled release of ampicillin. Int J Pharm 312:166–173
8. Kiilll CP et al (2017) Synthesis and factorial design applied to a novel chitosan/sodium polyphosphate nanoparticles via ionotropic gelation as an RGD delivery system. Carbohydr Polym 157:1695–1702
9. Amidi M, Mastrobattista E, Jiskoot W, Hennink WE (2010) Chitosan-based delivery systems for protein therapeutics and antigens. Adv Drug Deliv Rev 62(1):59–82
10. Luo Y, Wang Q (2014) Recent development of chitosan-based polyelectrolyte complexes with natural polysaccharides for drug delivery. Int J Biol Macromol 64:353–367
11. Sarika PR, James NR (2016) Polyelectrolyte complex nanoparticles from cationised gelatin and sodium alginate for curcumin delivery. Carbohydr Polym 148:354–361
12. Lankalapalli S, Kolapalli VRM (2019) Polyelectrolyte Complexes: A Review of their Applicability in Drug Delivery Technology. Indian J. Pharm. Sci. 71(5):481–487
13. Brovko OS, Palamarchuk IA, Val NA, Chukhchin DG (2017) Gels of Sodium Alginate–Chitosan Interpolyelectrolyte Complexes. Russ J Phys Chem A 91(8):1580–1585
14. Cerchiara T et al (2016) Microparticles based on chitosan/carboxymethylcellulose polyelectrolyte complexes for colon delivery of vancomycin. Carbohydr Polym 143:124–130
15. Le D, Sophie A, Momar P, Cristea M, Karakasyan-dia C, Picton L (2014) Formation of polyelectrolyte complexes with diethylaminoethyl dextran: Charge ratio and molar mass effect. Carbohydr Polym 113:217–224
16. Kumar A, Ahuja M (2013) Carboxymethyl gum kondagogu—chitosan polyelectrolyte complex nanoparticles: Preparation and characterization. Int J Biol Macromol 62:80–84
17. Polexe RC, Delair T (2013) Elaboration of stable and antibody functionalized positively charged colloids by polyelectrolyte complexation between chitosan and hyaluronic acid. Molecules 18:8563–8578
18. Tsai R et al (2014) Chitosan/pectin/gum Arabic polyelectrolyte complex: Process-dependent appearance, microstructure analysis and its application. Carbohydr Polym 101:752–759
19. Costalat M, Alcouffe P, David L, Delair T (2015) Macro-hydrogels versus nanoparticles by the controlled assembly of polysaccharides. Carbohydr Polym 134:541–546
20. Schatz C, Domard A, Viton C, Pichot C (2004) Versatile and Efficient Formation of Colloids of Biopolymer-Based Polyelectrolyte Complexes. Biomacromol 5:1882–1892
21. Bigucci F et al (2008) Chitosan/pectin polyelectrolyte complexes: Selection of suitable preparative conditions for colon-specific delivery of vancomycin. Eur J Pharm Sci 35:435–441
22. Sadat F, Tekie M, Kiani M, Zakerian A, Pilevarian F (2017) Nano polyelectrolyte complexes of carboxymethyl dextran and chitosan to improve chitosan-mediated delivery of miR-145. Carbohydr Polym 159:66–75

23. Wasupalli GK, Verma D (2018) Molecular interactions in self-assembled nano-structures of chitosan–sodium alginate based polyelectrolyte complexes. Int J Biol Macromol 114:10–17
24. Wu D, Delair T (2015) Stabilization of chitosan/hyaluronan colloidal polyelectrolyte complexes in physiological conditions. Carbohydr Polym 119:149–158
25. Voron NG, Derkach SR, Kuchina YA, Sokolan NI (2016) The chitosan–gelatin (bio) polyelectrolyte complexes formation in an acidic medium. Carbohydr Polym 138:265–272
26. Sarmento B, Ribeiro AJ, Veiga F, Ferreira DC, Neufeld RJ (2007) Insulin-Loaded Nanoparticles are Prepared by Alginate Ionotropic Pre-Gelation Followed by Chitosan Polyelectrolyte Complexation. J Nanosci Nanotechnol 7(8):2833–2841
27. Saladini B, Bigucci F, Cerchiara T, Gallucci MC, Luppi B (2012) Microparticles based on chitosan/pectin polyelectrolyte complexes for nasal delivery of tacrine hydrochloride. Drug Deliv Transl Res 3:33–41
28. Delair T (2011) Colloidal polyelectrolyte complexes of chitosan and dextran sulfate towards versatile nanocarriers of bioactive molecules. Eur J Pharm Biopharm 78(1):10–18
29. Madhusudana K, Kumar A, Soo S (2018) Polysaccharide-based magnetically responsive polyelectrolyte hydrogels for tissue engineering applications. J Mater Sci Technol 34(8):1371–1377
30. Tan C, Xie J, Zhang X, Cai J, Xia S (2016) Polysaccharide-based nanoparticles by chitosan and gum arabic polyelectrolyte complexation as carriers for curcumin. Food Hydrocoll 57:236–245
31. Lalevée G, Sudre G, Montembault A, Meadows J, Malaise S, Crépet A (2016) Polyelectrolyte complexes via desalting mixtures of hyaluronic acid and chitosan —Physicochemical study and structural analysis. Carbohydr Polym 154:86–95
32. Kulig D, Zimoch-Korzycka A, Jarmoluk A, Marycz K (2016) Study on Alginate–Chitosan Complex Formed with different polymer ratio. Polymers (Basel) 167:167–183
33. Feng Q, Zeng G, Yang P, Wang C, Cai J (2005) Self-assembly and characterization of polyelectrolyte complex films of hyaluronic acid/chitosan, 258:85–88
34. Quadrado RFN, Fajardo AR (2018) Microparticles based on carboxymethyl starch/chitosan polyelectrolyte complex as vehicles for drug delivery systems. Arab J Chem
35. Li X, Wu Z, Zhang B, Pan Y, Meng R, Chen H (2019) Fabrication of chitosan hydrochloride and carboxymethyl starch complex nanogels as potential delivery vehicles for curcumin. Food Chem 293(April):197–203
36. Henao E, Delgado E, Contreras H, Quintana G (2018) Polyelectrolyte complexation versus ionotropic gelation for chitosan-based hydrogels with carboxymethylcellulose, carboxymethyl starch, and alginic acid. Int J. Chem Eng 2018:1–12
37. Patil JS, Kamalapur MV, Marapur SC, Kadam DV (2010) Ionotropic gelation and polyelectrolyte complexation: the novel techniques to design hydrogel particulate sustained, modulated drug delivery system: a review. Dig J Nanomater Biostructures 5(1):241–248
38. Hamman JH (2014) Chitosan based polyelectrolyte complexes as potential carrier materials in drug delivery systems. Mar Drugs 8(June):1305–1322
39. Sadeghi AMM, Dorkoosh FA, Avadi MR, Saadat P, Rafiee-tehrani M, Junginger HE (2008) Preparation, characterization and antibacterial activities of chitosan, N -trimethyl chitosan (TMC) and N -diethylmethyl chitosan (DEMC) nanoparticles loaded with insulin using both the ionotropic gelation and polyelectrolyte complexation methods. Int J Pharm 355:299–306
40. Liu Z, Jiao Y, Wang Y, Zhou C, Zhang Z (2008) Polysaccharides-based nanoparticles as drug delivery systems. Adv Drug Deliv Rev 60(15):1650–1662
41. Kulig D, Zimoch-Korzycka A, Jarmoluk A, Marycz K (2016) Study on Alginate–Chitosan complex formed with. Polymers (Basel) 167:1–17
42. Sarmento B, Ferreira D, Veiga F, Ribeiro A (2006) Characterization of insulin-loaded alginate nanoparticles produced by ionotropic pre-gelation through DSC and FTIR studies. Carbohydr Polym 66:1–7
43. Hudson D, Margaritis A (2014) Biopolymer nanoparticle production for controlled release of biopharmaceuticals. Crit Rev Biotechnol 34(July):161–179
44. Gierszewska M, Ostrowska-czubenko J, Chrzanowska E (2018) pH-responsive chitosan/alginate polyelectrolyte complex membranes reinforced by tripolyphosphate. Eur Polym J 101:282–290

45. Volod AV et al (2014) Soluble chitosan–carrageenan polyelectrolyte complexes and their gastroprotective activity. Carbohydr Polym 101:1087–1093
46. Volod AV, Davydova VN, Glazunov VP, Likhatskaya GN, Yermak IM (2016) Influence of structural features of carrageenan on the formation of polyelectrolyte complexes with chitosan. Int J Biol Macromol 84:434–441
47. Vlaia L, Coneac G, Olariu I, Vlaia V, Lupuleasa D Cellulose-derivatives-based hydrogels as vehicles for dermal and transdermal drug delivery
48. Yan J, Wang Y, Qiu W, Wu J (2017) Construction and characterization of nanosized curdlan sulfate/chitosan polyelectrolyte complex toward drug release of zidovudine. Carbohydr Polym 174:209–216
49. Yan J, Qiu W, Wang Y, Wu L, Cheung PCK (2018) Formation and characterization of polyelectrolyte complex synthesized by chitosan and carboxylic curdlan for 5-fluorouracil delivery. Int J Biol Macromol 107:397–405
50. Krauland AH, Alonso MJ (2007) Chitosan/cyclodextrin nanoparticles as macromolecular drug delivery system. Int J Pharm 340:134–142
51. Silva DA, Maciel JS, Feitosa JPA, Paula HCB, De Paula RCM (2010) Polysaccharide-based nanoparticles formation by polyeletrolyte complexation of carboxymethylated cashew gum and chitosan. J Mater Sci 45:5605–5610
52. Wu D et al (2017) Ternary polysaccharide complexes: Colloidal drug delivery systems stabilized in physiological media. Carbohydr Polym 172:265–274
53. Giri TK et al (2019) Sustained release of diltiazem hydrochloride from cross-linked biodegradable IPN hydrogel Beads of pectin and modified xanthan. Indian J Pharm Sci 75(6):619–627
54. da Costa MPM, Ferreira ILM, Cruz MTM (2016) New polyelectrolyte complex from pectin/chitosan and montmorillonite clay. Carbohydr Polym 146:123–130
55. Drogoz A et al (2007) Polyelectrolyte Complexes from Polysaccharides: formation and Stoichiometry Monitoring. Langmuir 23(14):10950–10958
56. Verma VS, Sakure K, Badwaik HR (2017) Xanthan gum a versatile biopolymer: current status and future prospectus in hydro gel drug delivery. Curr Chem Biol 11:10–20
57. Kim J, Hwang J, Seo Y, Jo Y, Son J, Choi J (2017) Engineered chitosan—xanthan gum biopolymers effectively adhere to cells and readily release incorporated antiseptic molecules in a sustained manner. J Ind Eng Chem 46:68–79

Chapter 7
Ionically Gelled Polysaccharide-Based Interpenetrating Polymer Network Systems for Drug Delivery

Check for updates

Mohsen Khodadadi Yazdi, Mohammad Reza Ganjali, Morteza Rezapour, Payam Zarrintaj, Sajjad Habibzadeh, and Mohammad Reza Saeb

Abstract Interpenetrating polymer networks (IPNs) are a special type of polymer blends benefiting from combination of properties of two or more homo-/copolymers. Polysaccharide-based IPNs are promising materials for designing and manufacturing drug delivery and/or theranostic platforms. Ionically crosslinked polysaccharide-based IPNs in which polymer chains are crosslinked using non-toxic metal ions or low-molecular weight organic ions can be utilized to develop advanced delivery systems, which can disassemble in response to specific exogenous or endogenous stimuli such as pH, temperature, ion concentration, and redox species. Accordingly, such smart and stable delivery platforms are very interesting biomaterials to be loaded with various therapeutic agents. This chapter summarizes the application of these fascinating composite biomaterials in fabrication of delivery platforms.

Keywords Interpenetrating polymer networks · IPNs · Semi-IPNs · Polysaccharides · Ionic cross-linking · Drug delivery

---

M. K. Yazdi · M. R. Ganjali
Center of Excellence in Electrochemistry, School of Chemistry, College of Science, University of Tehran, Tehran, Iran

M. R. Ganjali (✉)
Biosensor Research Center, Endocrinology and Metabolism Molecular-Cellular Sciences Institute, Tehran University of Medical Sciences, Tehran, Iran
e-mail: ganjali@ut.ac.ir

M. Rezapour
Research Institute of Petroleum Industry (RIPI), P.O. Box 14665-137, Tehran, Iran

P. Zarrintaj
School of Chemical Engineering, Oklahoma State University, 420 Engineering North, Stillwater, OK 74078, USA

S. Habibzadeh
Department of Chemical Engineering, Amirkabir University of Technology (Tehran Polytechnic), Tehran, Iran

M. R. Saeb
Université de Lorraine, CentraleSupélec, LMOPS, F-57000, Metz, France

A. K. Nayak et al. (eds.), *Ionically Gelled Biopolysaccharide Based Systems in Drug Delivery*, Gels Horizons: From Science to Smart Materials,
https://doi.org/10.1007/978-981-16-2271-7_7

1 Introduction

Drugs, as therapeutic agents, are the indispensable body of medicine and medical science. According to the statistics, the global market for pharmaceuticals has reached $1.2 trillion just in 2018 [1]. Drug delivery systems (DDSs) have been in the core of attention in the last decades because of emerging novel therapeutics with complex and delicate chemical architecture and developing new diagnosis techniques that has revolutionized the medicine [2, 3]. Conventional DDSs mainly include local (e.g., topical) or systemic (e.g., oral delivery, intravenous injection, rectal administration, inhalation) delivery of therapeutics. Most of drugs are administered through the systemic delivery, which is then absorbed into blood flow to reach the proper location. However, the concentration of such therapeutic agents typically increases to a maximum value following by a decaying trend such that frequent controlling is essential to maintain the blood drug concentration at a minimum required level. Drug expenses, which is related to complex manufacturing methods, and adverse side effects (e.g., vomiting and hair loss for chemotherapeutics) have limited the conventional systemically administered drugs.

In the last two decades, there has been a global shift from drug administration with limited efficiency towards direct drug delivery in organ, tissue, or even cells [4]. Such methods preserve drug stability and bioactivity before reaching the region of action and reduce the required dose of drug, which is highly beneficial for unstable (e.g., proteins) or toxic therapeutic molecules (e.g., cancer drugs). DDSs that deliver the therapeutic agents to cancer cells while ignoring healthy ones are the final goal in targeted cancer therapy [5]. However, designing smart delivery vehicles that can detect cancer cells and transfer drugs into intracellular environment is quite challenging [6]. Furthermore, penetration of drug carriers into biological barriers such as intestinal mucosa and blood-brain barrier (BBB) should be considered as well [7]. On the other hand, evading the immune system is another important challenge when engineering DDSs [8]. It may be required to control drug delivery systems to penetrate across cells' membrane to deliver therapeutics (e.g., chemotherapeutics, proteins, RNA-based drugs) directly into the cytosol [9, 10].

Nowadays, various DDSs with different shapes, constructing materials, and size are used for designing DDSs, and a specific attention if focused on nanoscale DDSs [11]. Nanocarriers provide high surface area for drug molecules to be adsorbed or high volume for drug encapsulation; besides, they can be engineered to penetrate the biological barriers and cell membrane and evade the immune system.

2 Biomaterials for Making Drug Delivery Systems

Biomaterial designed for and implemented in medicine has been a necessity over the past century [12–14]. Various biomaterials, especially biopolymers, have been vastly utilized for fabrications of drug-carrying vehicles [15]. Versatile chemistry, a wide range physical and chemical properties, biocompatibility, and biodegradability have made biopolymers robust constructing materials for designing tailor-made delivery platforms [16–18]. There are various synthetic and natural-based biopolymers, which have been widely utilized in biomedical field [19–22]. Polysaccharides are naturally derived biopolymers containing disaccharide units with a wide range of biomedical applications [23–25]. Drug delivery systems based on polysaccharides or containing polysaccharides have gained many interests in the last decades [26].

These natural biopolymers benefit from biocompatibility, non-immunogenicity, and high water absorption [27–29]. There are many polysaccharides with various chemical structures and surface charges. Among different polysaccharides, chitosan is the most well-known polycationic polysaccharides while most of the famous polysaccharides (e.g., alginate) bear negative charges or are nearly neutral (e.g., agarose). Oppositely charged polysaccharides can make polyelectrolytes complexes (PEC) to make delivery nanoparticles. Besides, they are more likely could be crosslinked using metallic cations (e.g., Ca^{2+}, Fe^{3+}) or small negative ions (e.g., tripolyphosphate, TPP) [30–32]. These physical crosslinking can reversibly disassemble to its building blocks, contrary to covalent crosslinking. Physical crosslinking and hydrogen bonding play pivotal role in biological systems. Reversible physical crosslinking strategies provide a robust methodology to design and fabricate tailor-made delivery platforms for various applications. For example, designing polyethyleneglycol (PEG)-shielded DDS to evade immune system is a common method which can result in PEG dilemma [33]; accordingly, developing cleavable or detachable PEGylation are very important to suppress PEG dilemma and enhance drug delivery efficacy [34]. However, reversibly detachable physical bonds such as ionic crosslinking, which is cleaved in response to alternation in physicochemical properties of microenvironments, is preferred over chemical bonds which normally need high energy radiation to be cleaved.

3 Interpenetrating Polymer Networks (IPNs)

There are various methods to combine various monomers, oligomers, or polymers to create novel macrostructures [35, 36]. Mixing two or more polymers create a polymer blend which can be separated based on different physical properties such as density. In contrast, in copolymers different nanoscale block are covalently attached to create fascinating heterostructures [23, 37]. An interpenetrating polymer network (IPN), based on IUPAC, is defined as: “A polymer comprising two or more networks which are at least partially interlaced on a molecular scale but not covalently bonded to each other and cannot be separated unless chemical bonds are broken” [38].Topological interlocking is the main feature of IPNs where co-continuous polymer chains are engaged such that they cannot be separated without chemical bond scission [39]. Many researchers have used IPN and semi-IPN in biomedical applications [40]. Especially, the IPN seem interesting for designing novel DDSs [41].

4 Polysaccharide-Based IPNs for Drug Delivery

IPN structures can enhance the mechanical properties and physiochemical stability of individual polymers such that polymer-based IPNs have been utilized in various biomedical fields. Many polysaccharides-based IPN or semi-IPN have used in tissue engineering and DDSs [42]. Many of such delivery systems are based on hydrogels with various materials and sizes; for example, Zhao et al. made multi-functional IPN injectable hydrogels for delivery of both small molecules and proteins [43]. They used methacrylated alginate, poly(ethylene glycol) methacrylate, and Poly(N-isopropylacrylamide)(PNIPAAm) to fabricate thermo-/pH-responsive hydrogels [44].

These IPN or semi-IPN usually utilize chemicals to crosslink one or both polymer chains in order to increase mechanical integrity and chemical stability. Some crosslinkers make robust covalent bonds while others create physical bonds between adjacent chains. For example, dialdehydes create chemical bonds between polymer chains containing NH_2 pendant groups such as gelatin and chitosan while calcium ions physically attach segments of two adjacent polymer chains via robust electrostatic interactions [45–47]. Accordingly, in contrast to chemical crosslinks, physical crosslinks can be disassembled reversibly which indicates that both crosslinking strategies have their own benefits and disadvantages. In this section, some chemical crosslinked semi-IPN and IPN based DDSs are introduced first followed by ionic crosslinked systems.

Chitosan and its derivatives are widely utilized polysaccharides in biomedical applications [48]. This biocompatible polymer is a positively charged polysaccharide, which can dissolve, in acidic aqueous solution. Accordingly, DDSs based on chitosan are suspicious to dissolution in stomach environment limiting oral administration of chitosan-based DDSs. However, instability of chitosan-based hydrogels under acidic conditions can be enhanced through manufacturing chitosan-based IPNs [49]. Indeed, chitosan/chitosan derivatives-based IPNs with improved stability can be more useful in controlled release applications. For example, a semi-IPN hydrogel based on hydroxyethyl cellulose (HEC)-grafted-acrylic acid was utilized for controlled delivery of perindopril erbumine [50]. The semi-IPN hydrogel showed pH sensitivity for which release rate of acid labile drug is significantly lower under acidic environments (i.e., pH 1.2). The hydrogel protect the drug in acidic condition in the stomach while release the drug more rapidly under neutral pH in small intestine or colon [50]. On the other hand, IPN hydrogels based on chitosan and poly(2-hydroxyethyl methacrylate) were fabricated for delivery of quetiapine as the model drug [51]. Chitosan was crosslinked using glutaraldehyde while *N,N′*-methylenebis(acrylamide) (BIS) was utilized to crosslink poly(2-hydroxyethyl methacrylate). In fact, glutaraldehyde have been the most utilized crosslinker for chitosan [52, 53]; on the other hand, many chemical crosslinker such as N,N′-methylene-bis-acrylamide (MBA), poly(ethyleneglycol) diglycidyl ether (PEGDGE), diacrylated PEG (PEGDA), BIS, and genipin have used to fabricate various semi-IPN and IPNs containing polysaccharides [54–58].

Designing DDSs for controlled or targeted delivery of chemotherapeutic agent have been in the center of attention for decades because of harmful side effects of these drugs. A pH-sensitive semi-IPN hydrogel was made for controlled delivery of an anticancer drug, 5-fluorouracil (5-FU) [59]. Poly (acrylamide-co-acrylic acid) (Poly (AAm-co-AA)) was created within N-succinyl-chitosan (NSC) network which was crosslinked using glutaraldehyde. On the other hand, since the pH vary in different part of the body (e.g., throughout the gastric tract) and microenvironments (e.g., pH tumor microenvironment is slightly acidic) designing pH-responsive DDSs have grabbed much attentions between material scientists and engineers. Chaves et al. used semi-IPN hydrogels based on maleoyl-chitosan (a water-soluble chitosan derivative) for delivery of dapsone which is a bacteriostatic and antibacterial drug [60]. The pH-responsive semi-IPN enhances the drug solubility and release behavior of drug in a controlled manner. Furthermore, Abbasi et al. fabricated a pH-sensitive biodegradable hydrogel based on pectin copolymers for delivery of sulfasalazine in ulcerative colitis [61]. Pectin was grafted to PEG and methacrylic acid (MAA) resulting in pectin-g-(PEG-co-MAA) polymer (see Fig. 1). Drug release was accelerated at colonic pH (i.e., pH 7.4) compared to acidic pH. Furthermore, it was observed that natural microflora of the colon can degrade the hydrogel [61].

Semi-IPNs and IPNs containing both polysaccharides and other natural polymers, which benefit from the attractive properties of both biopolymers can be prepared for special requirements. For example, biodegradation of polysaccharide-based IPNs can be improved by making IPNs with biodegradable gelatin. A semi-IPNs based on gelatin/sodium carboxymethyl cellulose (NaCMC) and gelatin/sodium carboxymethyl nanocellulose (NaCMNC) were prepared for delivery of Tramadol

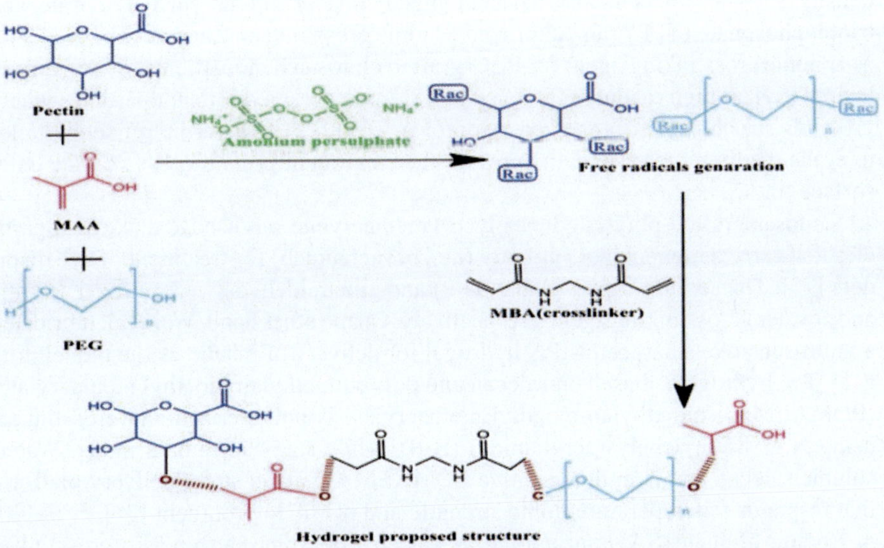

Fig. 1 The chemical structure proposed for pectin-g-(PEG-co-MAA) polymer [47]

[62]. Gelatin chains were crosslinked using the glutaraldehyde to make covalent bonds; drug encapsulation efficiency was higher for cellulose microsphere compared to nanocellulose while release rate was lower for nanocellulose microsphere [62].

Hyaluronic acid (HA) is an anionic biodegradable polysaccharide with extensive applications in tissue engineering and DDSs [63]. It is a major component of extracellular matrix (ECM) of connective tissue which present in body fluids, too. Cluster determinant 44 (CD44) receptors, which is a transmembrane glycoprotein that is overexpressed in many cancers, is the receptor for HA. Accordingly, HA-based or HA-modified carriers have been widely used in targeted cancer therapy [64, 65]. A semi-IPN hydrogel micro-particles (microgel) based on HA were fabricated an injectable platform for release of Rhodamine 6G as the model drug [66]. The microgels were modified using various polymers including polyvinylpyrrolidone (PVP), PEG, and PEG methyl ether (PEGME) to improve the biocompatibility and hemocompatibility whiled ivinyl sulfone (DVS) was used as crosslinker for HA. The presence of HA in semi-IPN microgel improve injection, dispersion, and drug release properties of these delivery systems. These microgels provide cost-effective platforms in tissue engineering and DDSs [66]. Another semi-IPN based on collagen and low-molecular-weight hyaluronic acid (LMW HA) was utilized for targeted delivery of chaperone protein Hsp70 for Parkinson's disease (PD) [67].

On the other hand, physical crosslinks are less robust compared to chemical crosslinks; however, there are beneficial aspects for such crosslinking as it provides fabrication of macrostructure that can be disassembled under certain conditions; because they normally utilize less toxic crosslinking agents. There are several physical crosslinking strategies such as hydrogen bonding, metal chelation, hydrophobic interactions, and ionic crosslinking [68]. Chemical species bearing high charge density such as divalent and trivalent metal ions or anions such as citrate and tripolyphosphate (TPP) can be utilized as ionic crosslinkers. Zhang et al., used TPP as an ionic crosslinking agent for PEG-grafted chitosan nanoparticles for delivery of insulin [69]. A high loading efficiency of 38% was obtained for nanoparticles where protein's structure was preserved. Moreover, stealth PEG layer help nanoparticles to evade the immune system and prevent the creation of protein corona on the outer surface [69].

Chitosan grafted-poly(ethylene glycol) methacrylate was used to make nanoparticles that carry a monoclonal antibody (i.e., bevacizumab) for treatment of eye disorders [70]. The sodium triphosphate (TPP) and glutaraldehyde were utilized as ionic and covalent crosslinking agents, respectively. On the other hand, Wei et al. fabricated a multi-sensitive smart semi-IPN hydrogel for delivery of insulin as the model drug [71]. The hydrogel is based on salecan and poly (dimethylaminoethyl methacrylate) (PDMAEMA); dimethylaminoethyl methacrylate is polymerized and crosslinked, using N,N′-Methylenebis(acrylamide) (BIS), while sale can remain intact. Water-soluble salecan and is multi-sensitive PDMAEMA make a smart delivery platform that responds to temperature, ionic strength, and pH of environment [71].

Sodium alginate (SA) is an anionic polysaccharide which can be easily crosslinked with divalent ions, especially calcium ions. Accordingly, SA is one of the most utilized polysaccharide for making ionically gelled polysaccharide-based IPNs. For example, Hu et al., manufactured a semi-IPN hydrogel based on hydroxypropyl methyl cellulose (HPMC) and SA for sustained delivery of both small-molecule drugs (metformin hydrochloride and indomethacin) and proteins (bovine serum albumin, BSA) [72]. The SA was ionically crosslinked with calcium ions, as shown in Fig. 2. They also coated hydrogel with chitosan in which electrostatic interactions between anionic SA and cationic chitosan create a PEC; in other words, dual physical crosslinking was used to fabricate this hydrogel. The resulting HPMC-SA semi-IPN are degradable with a good water uptake capability and higher drug loading capacity. However, small molecule drugs was released with a burst release behavior, which was related to large pore size of HPMC-SA hydrogels, contrary to macromolecular drug [72].

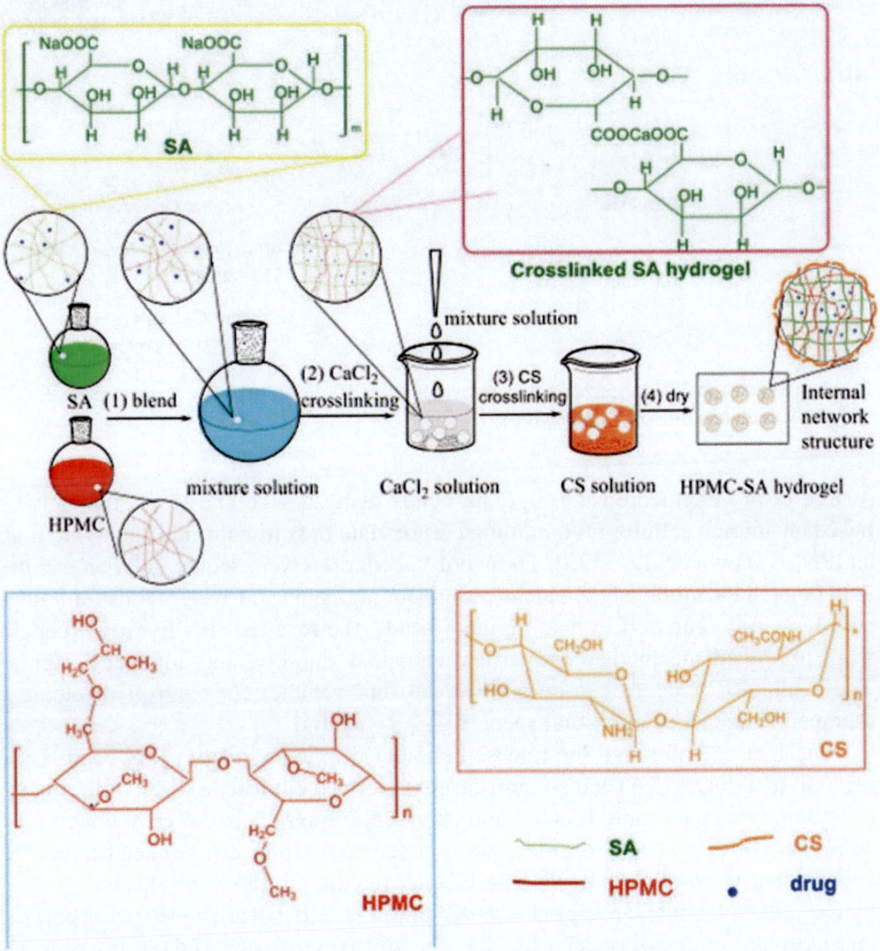

Fig. 2 Schematic representation of production process of HPMC-SA hydrogels and chemical structure of the hydrogels [58]

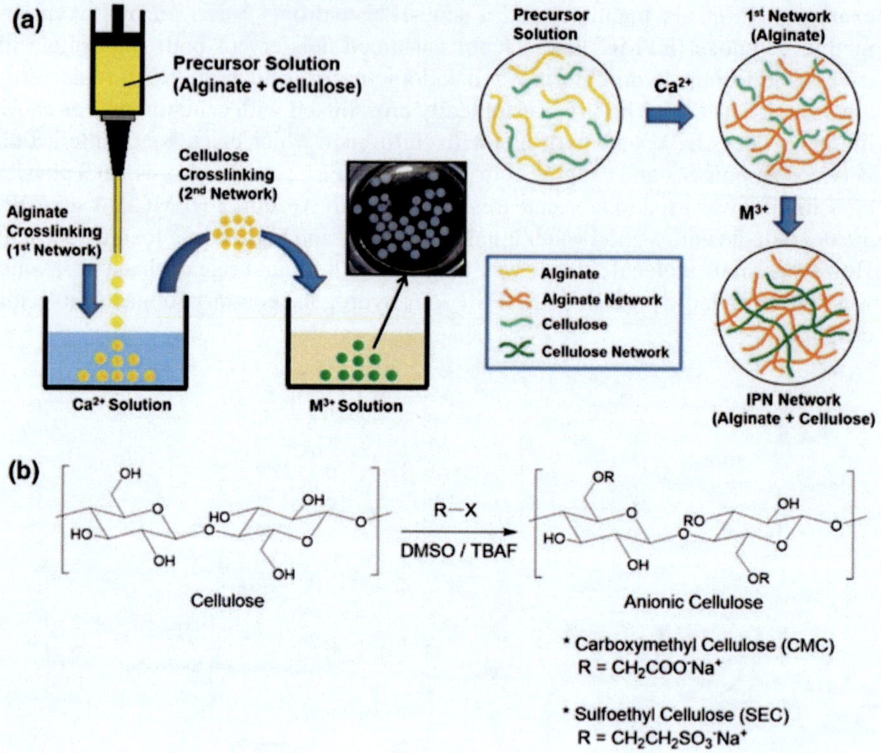

Fig. 3 **a** Schematic illustration of dual ionic crosslinking IPN beads based on alginate and cellulose. **b** Esterification reaction of cellulose [60]

Lee et al., engineered the alginate beads using a secondary polymer network based on anionic cellulose (containing carboxylate or sulfonate side chains) to make an IPN, as shown in Fig. 3 [73]. These polysaccharides were sequentially crosslinked using metallic cations, i.e., divalent cations for alginate and trivalent ions for anionic cellulose. As compared to neat alginate beads, the resulted IPN hydrogel benefits from improved mechanical properties, chemical stability, and adjustable rigidity. The resulted IPN beads can be used as delivery vehicles for controlled release of therapeutic agents or microbial species [73].

Rajesh et al. compared the release behavior of ofloxacin from IPN beads based on sodium alginate (SA) and sodium carboxymethyl cellulose (SCMC) utilizing two crosslinking agents: ionic ($FeCl_3$) and chemical (glutaraldehyde) crosslinkers [74]. A Fickian diffusion behavior of drug was observed for ionic crosslinked beads while for covalently crosslinked beads a zero-order release profiles was detected.

Regardless of chitosan, alginate, and hyaluronic acid that are widely used polysaccharide, many other polysaccharides have been also used to make DDSs. For example, Lohani et al., prepared IPN pH-responsive beads using two polysaccharide namely k-Carrageenan and sodium carboxymethyl cellulose (SCMC) for delivery of Ibuprofen

[75]. The beads were crosslinked using $AlCl_3$ which is the source for trivalent Al^{3+} ions. It was observed that drug release is significantly higher at higher pH, i.e., pH 7.4, compared to acidic pH (i.e., pH 1.2). An anomalous drug release behavior was detected for the IPN beads. The obtained results indicate that this formulation can be utilized for oral delivery of therapeutic agent where they are protected in highly acidic environment in the stomach while release at nearly neutral pH of the small intestine or colon [75].

Controlled release of bovine serum albumin (BSA) and diclofenac sodium from polysaccharide-based IPN hydrogel of gellan and xanthan was studied [76]. The pH-sensitive hydrogel is embedded with polymethyl methacrylate (PMMA) nanoparticles to adjust release behavior and mechanical properties of nerve conduits. The gel is crosslinked by heating and calcium ions to make robust hydrogel conduits. A zero-order release behavior was observed for both drugs during 20 and 30 days, for BSA and diclofenac sodium, respectively.

As discussed in this section, both ionic and chemical crosslinking agents have been widely utilized to fabricate polysaccharide-based IPN DDSs. However, ionically gelled polysaccharide-based IPNs which better mimic many of natural macrostructures (e.g., DNA) will experience ever-growing applications in novel DDSs.

5 Conclusion

Polysaccharide-based IPNs are valuable polymeric structures for tissue engineering and drug delivery applications. Ionic crosslinking agents in which high charge density molecules are utilized to interact with charged polysaccharides provide the user with a robust strategy for making reversible IPN macrostructures. Ionic crosslinking improves the mechanical properties, chemical stability, and water retention features of the IPN hydrogels. Accordingly, they improve drug loading capacity and release behavior such that more drugs can be entrapped into swollen porous network of IPN hydrogel. Such systems can be designed to respond to microenvironmental conditions in target regions in the human body and rearrange to create new structures while discarding no or minor toxic species. Flexibility in design of complex macrostructures for ionically gelled polysaccharide-based IPNs for advanced DDSs would enhance the applications of these systems in the near future.

References

1. I. I. f. H. D. Science (2019) The global use of medicine in 2019 and outlook to 2023
2. Allen TM, Cullis PR (2004) Drug delivery systems: entering the mainstream. Science 303(5665):1818–1822

3. Rabiee N, Bagherzadeh M, Jouyandeh M, Zarrintaj P, Saeb MR, Mozafari M, Shokouhimehr M, Varma RS (2021) Natural Polymers Decorated MOF-MXene Nanocarriers for Co-delivery of Doxorubicin/pCRISPR ACS Appl. Bio Mater. https://doi.org/10.1021/acsabm.1c00332
4. Öztürk-Atar K, Eroğlu H, Çalış S (2018) Novel advances in targeted drug delivery. J Drug Target 26(8):633–642
5. Kato T (2019) Encapsulated drugs in targeted cancer therapy. Controlled drug delivery 11
6. Rahmati M, Mills SK, Urbanska AM, Saeb MR, Venugopal JR, Ramakrishna S, Mozafari M (2021) Electrospinning for tissue engineering applications. Prog Mater Sci 117:100721
7. Laksitorini M, Prasasty VD, Kiptoo PK, Siahaan TJ (2014) Pathways and progress in improving drug delivery through the intestinal mucosa and blood–brain barriers. Therapeutic Deliv 5(10):1143–1163
8. Fadeel B (2019) Hide and seek: nanomaterial interactions with the immune system. Frontiers in immunology 10
9. Lee Y-W, Luther DC, Kretzmann JA, Burden A, Jeon T, Zhai S, Rotello VM (2019) Protein delivery into the cell cytosol using non-viral nanocarriers. Theranostics 9(11):3280
10. Zhu Q, Qi H, Long Z, Liu S, Huang Z, Zhang J, Wang C, Dong L (2016) Extracellular control of intracellular drug release for enhanced safety of anti-cancer chemotherapy. Scient Rep 6(1):1–11
11. Majumder J, Taratula O, Minko T (2019) Nanocarrier-based systems for targeted and site specific therapeutic delivery. Adv Drug Deliv Rev 144:57–77
12. Zarrintaj P, Bakhshandeh B, Rezaeian I, Heshmatian B, Ganjali MR (2017) A novel electroactive agarose-aniline pentamer platform as a potential candidate for neural tissue engineering. Scient Rep 7(1):1–12
13. Zarrintaj P, Bakhshandeh B, Saeb MR, Sefat F, Rezaeian I, Ganjali MR, Ramakrishna S, Mozafari M (2018) Oligoaniline-based conductive biomaterials for tissue engineering. Acta Biomater 72:16–34
14. Zarrintaj P, Moghaddam AS, Manouchehri S, Atoufi Z, Amiri A, Amirkhani MA, Nilforoushzadeh MA, Saeb MR, Hamblin MR, Mozafari M (2017) Can regenerative medicine and nanotechnology combine to heal wounds? The search for the ideal wound dressing. Nanomedicine 12(19):2403–2422
15. Pappalardo D, Mathisen T, Finne-Wistrand A (2019) Biocompatibility of resorbable polymers: a historical perspective and framework for the future. Biomacromolecules, 20(4):1465–1477
16. Bagheri B, Zarrintaj P, Surwase SS, Baheiraei N, Saeb MR, Mozafari M, Kim YC, Park OO (2019) Self-gelling electroactive hydrogels based on chitosan–aniline oligomers/agarose for neural tissue engineering with on-demand drug release. Colloids Surf, B 184:
17. Shojaie S, Rostamian M, Samadi A, Alvani MAS, Khonakdar HA, Goodarzi V, Zarrintaj R, Servatan M, Asefnejad A, Baheiraei N (2019) Electrospun electroactive nanofibers of gelatin-oligoaniline/Poly (vinyl alcohol) templates for architecting of cardiac tissue with on-demand drug release. Polym Adv Technol 30(6):1473–1483
18. Servatan M, Zarrintaj P, Mahmodi G, Kim S-J, Ganjali MR, Saeb MR, Mozafari M (2020) Zeolites in drug delivery: progress, challenges and opportunities. Drug Discovery Today
19. Verma D, Gulati N, Kaul S, Mukherjee S, Nagaich U (2018) Protein based nanostructures for drug delivery. J Pharm
20. Gopinath V, Saravanan S, Al-Maleki A, Ramesh M, Vadivelu J (2018) A review of natural polysaccharides for drug delivery applications: Special focus on cellulose, starch and glycogen. Biomed Pharmacother 107:96–108
21. Mir M, Ahmed N, Rehman A (2017) Recent applications of PLGA based nanostructures in drug delivery. Colloids SurfS B: Biointerfaces 159:217–231
22. Schneider C, Langer R, Loveday D, Hair D (2017) Applications of ethylene vinyl acetate copolymers (EVA) in drug delivery systems. J Controlled Release 262:284–295
23. Zarrintaj P, Ahmadi Z, Saeb MR, Mozafari M (2018) Poloxamer-based stimuli-responsive biomaterials. Mater Today: Proc 5(7):15516–15523
24. Seidi F, Khodadadi Yazdi M, Jouyandeh M, Dominic M, Naeim H, Nasirnezhad M, Bagheri B, Habibzadeh S, Zarrintaj P, Saeb MR, Mozafari M (2021) Chitosan-based blends for biomedical applications. Int J Biol Macromol 183:1818–1850

25. Khosravi A, Fereidoon A, Khorasani MM, Naderi G, Ganjali MR, Zarrintaj P, Saeb MR, Gutiérrez TJ (2020) Soft and hard sections from cellulose-reinforced poly(lactic acid)-based food packaging films: A critical review. Food Packag Shelf Life 23:100429
26. Shokri Z, Seidi F, Karami S, Li C, Saeb MR, Xiao H (2021) Laccase immobilization onto natural polysaccharides for biosensing and biodegradation. Carbohyd Polym 262:117963
27. Mohebbi S, Nezhad MN, Zarrintaj P, Jafari SH, Gholizadeh SS, Saeb MR, Mozafari M (2019) Chitosan in biomedical engineering: a critical review. Curr Stem Cell Res Ther 14(2):93–116
28. Alizadeh R, Zarrintaj P, Kamrava SK, Bagher Z, Farhadi M, Heidari F, Komeili A, Gutiérrez TJ, Saeb MR (2019) Conductive hydrogels based on agarose/alginate/chitosan for neural disorder therapy. Carbohyd Polym 224:
29. Zarrintaj P, Saeb MR, Ramakrishna S, Mozafari M (2018) Biomaterials selection for neuroprosthetics. Curr Opin Biomed Eng 6:99–109
30. Patel N, Lalwani D, Gollmer S, Injeti E, Sari Y, Nesamony J (2016) Development and evaluation of a calcium alginate based oral ceftriaxone sodium formulation. Prog Biomater 5(2):117–133
31. Karimi M, Avci P, Ahi M, Gazori T, Hamblin MR, Naderi-Manesh H (2013) Evaluation of chitosan-tripolyphosphate nanoparticles as a p-shRNA delivery vector: formulation, optimization and cellular uptake study. J Nanopharmaceutics Drug Deliv 1(3):266–278
32. Boddohi S, Moore N, Johnson PA, Kipper MJ (2009) Polysaccharide-based polyelectrolyte complex nanoparticles from chitosan, heparin, and hyaluronan. Biomacromol 10(6):1402–1409
33. Hatakeyama H, Akita H, Harashima H (2013) The polyethyleneglycol dilemma: advantage and disadvantage of PEGylation of liposomes for systemic genes and nucleic acids delivery to tumors. Biol Pharm Bull 36(6):892–899
34. Fang Y, Xue J, Gao S, Lu A, Yang D, Jiang H, He Y, Shi K (2017) Cleavable PEGylation: a strategy for overcoming the "PEG dilemma" in efficient drug delivery. Drug Deliv 24(2):22–32
35. Mostafapoor F, Khosravi A, Fereidoon A, Khalili R, Jafari SH, Vahabi H, Formela K, Saeb MR (2020) Interface analysis of compatibilized polymer blends. In: Compatibilization of polymer blends. Elsevier, pp 349–371
36. Zarrintaj P, Saeb MR, Jafari SH, Mozafari M (2020) Application of compatibilized polymer blends in biomedical fields. In: Compatibilization of polymer blends. Elsevier, pp 511–537
37. Rohani Rad H, Vahabi K, Formela MR, Saeb MR, Thomas S (2019) Injectable poloxamer/graphene oxide hydrogels with well-controlled mechanical and rheological properties. Polym Adv Technol 30(9):2250–2260
38. McNaught AD, Wilkinson A (1997) Compendium of chemical terminology. Blackwell Science, Oxford
39. Myung D, Waters D, Wiseman M, Duhamel PE, Noolandi J, Ta CN, Frank CW (2008) Progress in the development of interpenetrating polymer network hydrogels. Polym Adv Technol 19(6):647–657
40. Qadri MF, Malviya R, Sharma PK (2015) Biomedical applications of interpenetrating polymer network system. Open Pharm Sci J 2(1)
41. Raina N, Rani R, Khan A, Nagpal K, Gupta M (2019) Interpenetrating polymer network as a pioneer drug delivery system: a review. *Polym Bull*, 1–24
42. Matricardi P, Di Meo C, Coviello T, Hennink WE, Alhaique F (2013) Interpenetrating polymer networks polysaccharide hydrogels for drug delivery and tissue engineering. Adv Drug Deliv Rev 65(9):1172–1187
43. Zhao J, Zhao X, Guo B, Ma PX (2014) Multifunctional interpenetrating polymer network hydrogels based on methacrylated alginate for the delivery of small molecule drugs and sustained release of protein. Biomacromol 15(9):3246–3252
44. Zarrintaj P, Jouyandeh M, Ganjali MR, Hadavand BS, Mozafari M, Sheiko SS, Russell AJ, Vatankhah-Varnoosfaderani M, Gutiérrez TJ, Saeb MR (2019) Thermo-sensitive polymers in medicine: A review. Eur Polym J
45. Kulkarni V, Kulkarni P, Keshavayya J (2007) Glutaraldehyde-crosslinked chitosan beads for controlled release of diclofenac sodium. J Appl Polym Sci 103(1):211–217
46. Campiglio CE, Contessi Negrini N, Farè S, Draghi L (2019) Cross-linking strategies for electrospun gelatin scaffolds. Materials 12(15):2476

47. Russo R, Malinconico M, Santagata G (2007) Effect of cross-linking with calcium ions on the physical properties of alginate films. Biomacromol 8(10):3193–3197
48. Manzoor K, Ahmad S, Soundarajan A, Ikram S, Ahmed S (2018) Chitosan based nanomaterials for biomedical applications. In: Handbook of nanomaterials for industrial applications. Elsevier, pp 543–562
49. Pal D, Nayak A, Saha S (2018) Interpenetrating polymer network hydrogels of chitosan: applications in controlling drug release. Cellulose-Based Superabsorbent Hydrogels, pp 1–41
50. Ijaz H, Tulain UR (2019) Development of interpenetrating polymeric network for controlled drug delivery and its evaluation. Int J Polym Mater Polym Biomater 68(18):1099–1107
51. García J, Ruiz-Durántez E, Valderruten N (2017) Interpenetrating polymer networks hydrogels of chitosan and poly (2-hydroxyethyl methacrylate) for controlled release of quetiapine. React Funct Polym 117:52–59
52. Krishnat K, Dhawale SC, Remeth JD, Havaldar VD, Kavitake PR (2017) Interpenetrating networks of carboxymethyl tamarind gum and chitosan for sustained delivery of aceclofenac. Marmara Pharm J 21(4):771–782
53. Kondolot Solak E, Er A (2016) pH-sensitive interpenetrating polymer network microspheres of poly (vinyl alcohol) and carboxymethyl cellulose for controlled release of the nonsteroidal anti-inflammatory drug ketorolac tromethamine. Artif Cells, Nanomedicine, Biotechnol 44(3):817–824
54. Li X, Li Q, Xu X, Su Y, Yue Q, Gao B (2016) Characterization, swelling and slow-release properties of a new controlled release fertilizer based on wheat straw cellulose hydrogel. J Taiwan Inst Chem Eng 60:564–572
55. Cui L, Jia J, Guo Y, Liu Y, Zhu P (2014) Preparation and characterization of IPN hydrogels composed of chitosan and gelatin cross-linked by genipin. Carbohyd Polym 99:31–38
56. Dragan ES, Cocarta AI, Gierszewska M (2016) Designing novel macroporous composite hydrogels based on methacrylic acid copolymers and chitosan and in vitro assessment of lysozyme controlled delivery. Colloids Surf, B 139:33–41
57. Liu XJ, Li HQ, Zhang BY, Wang YJ, Ren XY, Guan S, Gao GH (2016) Highly stretchable and tough pH-sensitive hydrogels with reversible swelling and recoverable deformation. Rsc Adv 6(6):4850–4857
58. Qi X, Wei W, Li J, Liu Y, Hu X, Zhang J, Bi L, Dong W (2015) Fabrication and characterization of a novel anticancer drug delivery system: salecan/poly (methacrylic acid) semi-interpenetrating polymer network hydrogel. ACS Biomater Sci & Eng 1(12):1287–1299
59. Bashir S, Teo YY, Naeem S, Ramesh S, Ramesh K (2017) pH responsive N-succinyl chitosan/Poly (acrylamide-co-acrylic acid) hydrogels and in vitro release of 5-fluorouracil. *PLoS One* 12(7)
60. Chaves LL, Silveri A, Vieira AC, Ferreira D, Cristiano MC, Paolino D, Di Marzio L, Lima SC, Reis S, Sarmento B (2019) pH-responsive chitosan based hydrogels affect the release of dapsone: Design, set-up, and physicochemical characterization. Int J Biol Macromol 133:1268–1279
61. Abbasi M, Sohail M, Minhas MU, Khan S, Hussain Z, Mahmood A, Shah SA, Kousar M (2019) Novel biodegradable pH-sensitive hydrogels: An efficient controlled release system to manage ulcerative colitis. Int J Biol Macromol 136:83–96
62. Kadry G (2019) Comparison between gelatin/carboxymethyl cellulose and gelatin/carboxymethyl nanocellulose in tramadol drug loaded capsule. Heliyon 5(9):
63. Prajapati VD, Maheriya PM (2019) Hyaluronic acid as potential carrier in biomedical and drug delivery applications. In: Functional polysaccharides for biomedical applications. Elsevier, pp 213–265
64. Luo Z, Dai Y, Gao H (2019) Development and application of hyaluronic acid in tumor targeting drug delivery. Acta Pharmaceutica Sinica B
65. Huang G, Huang H (2018) Hyaluronic acid-based biopharmaceutical delivery and tumor-targeted drug delivery system. J Controlled Release 278:122–126
66. Chen Q, Passos A, Balabani S, Chivu A, Zhao S, Azevedo HS, Butler P, Song W (2018) Semi-interpenetrating network hyaluronic acid microgel delivery systems in micro-flow. J Colloid Interface Sci 519:174–185

67. Tunesi M, Raimondi I, Russo T, Colombo L, Micotti E, Brandi E, Cappelletti P, Cigada A, Negro A, Ambrosio L (2019) Hydrogel-based delivery of Tat-fused protein Hsp70 protects dopaminergic cells in vitro and in a mouse model of Parkinson's disease. NPG Asia Mat 11(1):1–15
68. Hu W, Wang Z, Xiao Y, Zhang S, Wang J (2019) Advances in crosslinking strategies of biomedical hydrogels. Biomaterials Sci 7(3):843–855
69. Zhang X, Teng D, Wu Z, Wang X, Wang Z, Yu D, Li C (2008) PEG-grafted chitosan nanoparticles as an injectable carrier for sustained protein release. J Mater Sci Mater Med 19(12):3525–3533
70. Savin C-L, Popa M, Delaite C, Costuleanu M, Costin D, Peptu CA (2019) Chitosan grafted-poly (ethylene glycol) methacrylate nanoparticles as carrier for controlled release of bevacizumab. Mater Sci Eng, C 98:843–860
71. Wei W, Li J, Qi X, Zhong Y, Zuo G, Pan X, Su T, Zhang J, Dong W (2017) Synthesis and characterization of a multi-sensitive polysaccharide hydrogel for drug delivery. Carbohyd Polym 177:275–283
72. Hu Y, Zhang S, Han D, Ding Z, Zeng S, Xiao X (2018) Construction and evaluation of the hydroxypropyl methyl cellulose-sodium alginate composite hydrogel system for sustained drug release. J Polym Res 25(7):148
73. Lee K, Hong J, Roh HJ, Kim SH, Lee H, Lee SK, Cha C (2017) Dual ionic crosslinked interpenetrating network of alginate-cellulose beads with enhanced mechanical properties for biocompatible encapsulation. Cellulose 24(11):4963–4979
74. Rajesh AM, Popat KM (2017) In vivo and in vitro taste masking of ofloxacin and sustained release by forming interpenetrating polymer network beads. Pharm Dev Technol 22(1):26–34
75. Lohani A, Singh G, Bhattacharya SS, Hegde RR, Verma A (2016) Tailored-interpenetrating polymer network beads of κ-carrageenan and sodium carboxymethyl cellulose for controlled drug delivery. J Drug Deliv Sci Technol 31:53–64
76. Ramburrun P, Kumar P, Choonara YE, du Toit LC, Pillay V (2017) Design and characterization of neurodurable gellan-xanthan pH-responsive hydrogels for controlled drug delivery. Expert Opin Drug Deliv 14(3):291–306

Chapter 8
Ionically Gelled Polysaccharide-Based Multiple-Units in Drug Delivery

M. D. Figueroa-Pizano and E. Carvajal-Millan

Abstract The design, development, and production of orally administrable multiple-unit dosage forms have mostly been based on the usage of natural polymers. Notably, the large group of polysaccharides has been taken into account by researches to produce multiple-unit dosage forms for drug administration because of their exceptional properties. Polysaccharides are biocompatible and biodegradable because they are similar to the body tissues and can be degraded by enzymatic processes. Currently, the use of ionic polysaccharides for assembly any type of particle for drug delivery systems is a trend. This chapter deals with the different polysaccharide-based Ionically gelled multiple-units in drug delivery.

Keywords Ionically gelled · Multiple-units · Drug delivery · Polysaccharide

1 Multiple-Units as Drug Delivery Systems

Nowadays, oral drug administration is the most used way by patients because it is a friendly, painless, and safe medication route [1–6]. The oral administration form offers several advantages over other routes, such as easy administration, patient compliance, and fabrication of solid formulations. Moreover, along the intestinal tract, the drug has a large surface area to be absorbed, and the mucosal layer could help its attachment [5, 6]. However, the design and development of orally administered drugs is not easy, since different aspects (technical and biological) must be considered to bring the drug to the place of interest [5, 6]. The main biological drawbacks that oral drugs have to face along the gastrointestinal tract (GIT) to reach the bloodstream and perform its effect are the acidic pH in the stomach and the enzymatic action [1, 5, 6]. Recent researches are focusing on the production of devices which protect and improve the pharmaceutical dosages of existing drugs to prevent the drug damage or degradation at the GIT [2]. Thus, the obtained results have allowed moving from the

M. D. Figueroa-Pizano (✉) · E. Carvajal-Millan (✉)
Research Center for Food and Development (CIAD), Carretera Gustavo Enrique Astiazarán Rosas No. 46. Col. La Victoria 83304, Hermosillo, Sonora, Mexico
e-mail: ecarvajal@ciad.mx

A. K. Nayak et al. (eds.), *Ionically Gelled Biopolysaccharide Based Systems in Drug Delivery*, Gels Horizons: From Science to Smart Materials,
https://doi.org/10.1007/978-981-16-2271-7_8

traditional dosage form to novel oral drug delivery systems, which are more effective [7].

Currently, the solid dosage forms (most used forms for oral administration) [3, 8] are categorizing into two general groups: the Single-Unit dosage forms and the Multiple-Unit dosage forms [8–12] (Fig. 1). The former category, also named as monolithic system, is a no-divided formulation that contains the active drug disseminated in the excipient material matrix [11–13]. After oral administration, the drug is released instantly from the matrix by simple diffusion, which is considering as immediate-release dosage form. In consequence, the plasma-drug concentration had large fluctuations, and dose dumping is presented [14]. Conversely, the Multiple-Unit dosage forms consist of minitablets or mini-units inside a larger tablet o capsule. They are divided formulations and are known as multilayer systems [2, 9, 12, 15, 16]. The Multiple-Unit dosage term was defined in the early 1950s and involved mini-particles as minitablets, pellets, and granules [7, 11, 16]. More recently, the term incorporates nano- or micro-particles and beads using as drug-releasing systems [2, 12, 15]. The multiple-unit dosage forms are divided after administration without compromise the individual characteristics of sub-units, allowing their distribution throughout the GIT and, regular emptying occurs avoiding dose dumping effect [8, 11, 15]. This kind of dosage forms offers others numerous advantages compared to the Single-Unit dosage form, and they become popular systems for oral drug administration (Fig. 2).

Multiple-unit dosage forms protect the drug against physical (acidic pH), chemical (enzymatic action), and mechanical (shear stresses produced by flowing gastric juices) damage along the GIT [4, 6, 11]. In that way, the drug maintains its properties and achieves the desired effect at the target site. Moreover, these divided oral dosage forms avoid the physicochemical interaction between the molecules of drugs or with their excipient [9]. Indeed, they represent an ideal form to encapsulate different active pharmaceutical compounds or drugs that are incompatible, as well as hydrophobic or with poor stability, because they are separating by a matrix [4, 8, 17]. Their fragmentation in several particles helps to disseminate the drug throughout the GIT, which results in better bioavailability and absorption [4, 18]. Also, the drug levels in plasma are maintaining as required trough time [8, 16]. Multiple-unit dosage forms offer an excellent platform to fabricate modified-release dosage forms and provide many pharmacokinetics and pharmacodynamics advantages [4, 12, 16]. Thus, oral drug can be designed to delay the drug-release time after administration (delayed-release form), or to release the drug at determined rate for a specific time (sustained-release form, as

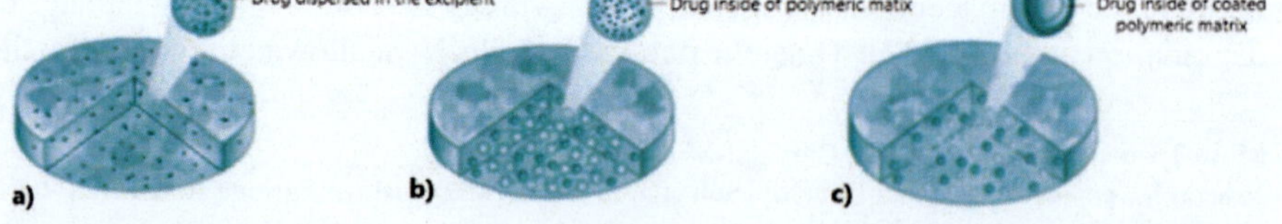

Fig. 1 Types of dosages forms: **a** Single-unit and **b** and **c** Multiple-units. *Source* Adapted from Hong Ding [14]

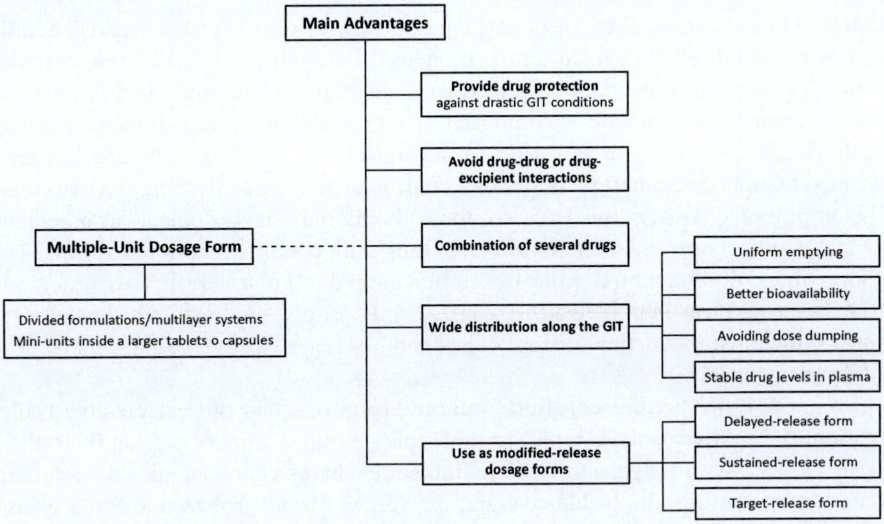

Fig. 2 Main advantages of multiple-unit dosage forms

part of extended-release form) with the intention to preserve the drug concentration in a constant manner as well as to release the drug at specific site in the body (site of action) or very close to it (target-release form) [4, 10, 14]. The final effects of this system have less variance in drug absorption, inter and intrapatient, and in time, so fewer doses application are needing by patients [2, 4, 7, 11].

2 Ionic Gelation of Polysaccharides for Production of Multiple-Unit Drug Delivery Systems

The design, development, and production of orally administrable multiple-unit dosage form have mostly been based on the usage of natural polymers. Over the years, natural polymers have been only used as excipients of pharmaceutical formulations to obtain the appropriate physical characteristics. However, they have the potential to improve the therapeutic effect of orally administrated dosages forms [3, 15]. So, the current trend is its use to formulate controlled drug delivery systems. Notably, the large group of polysaccharides has been taken into account by researches to produce multiple-unit dosage forms for drug administration because of their exceptional properties. In principle, they are the most abundant biopolymers in nature, so they are readily available, and in some cases, can be obtained from secondary or waste products. This fact makes them a low-cost material [19–21]. The principal sources of polysaccharides are plants, seaweeds, microorganisms, insects, crustaceans, and human tissues. Polysaccharides are biocompatible and biodegradable because they are similar to the body tissues and can be degraded by enzymatic via [19, 21–24].

Most of them have excellent mechanical properties, are easy to process, and generally are water-soluble [21, 25]. In terms of chemical structure, the distinctive polysaccharides properties are highly defined by their composition and chain conformations. Their basic structure is composed of monosaccharide units linked by glucosidic bonds. When monosaccharide units are the same in all the structure they have named homopolysaccharide; if monosaccharideare units are different they are called heteropolysaccharide [19–24]. Also, their chains may have a linear arrangement, while in other cases, their chains are branching with pending functional groups [24]. Sometimes, the functional groups become charged at polar solvents, as water, and the molecule turns in polyelectrolyte [26, 27]. Precisely, this feature has resulted in one of the most interesting for researches and has served as the basis for many drug delivery systems.

Ionic polysaccharides constitute a large group that can present positive (polycationic), negative (poly-anionic) or both charges (poly-ampholytic) on their chain [26, 28, 29]. As a polyelectrolyte, the molecule shares characteristics of polymers because of the high molecular weight, as well as an electrolyte due to its charge [27]. The ionization of the polysaccharide chain is due to the adsorption of ions from the solution or due to the dissociation of counter ions from the molecule [26]. However, the final charge in the polymer is determined by the density and distribution of ionizable groups along with it [29]. The ionization allows the chain expansion by repulsion between the charged groups, and also the free counter ions surround the chain creating an ionic environment (Fig. 3) [28]. Under these conditions, polysaccharides are water-soluble, present high swelling capacity, and are highly sensitives to pH changes [19, 30]. These new features are considered for the design of better drug delivery systems with pH or thermosensitivity. They can also interact with oppositely charged small molecules o other PEL to form different structures with particular properties [21, 26, 28].

Currently, the use of ionic polysaccharides for assembly of any type of particle for drug delivery systems is a trend. For ionotropic gelation chitosan (polycationic molecule) and alginate, carrageenan, pectin, xanthan gum, hyaluronan, heparin,

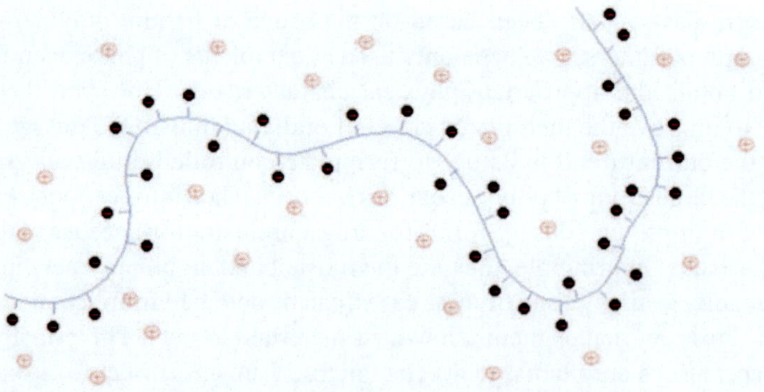

Fig. 3 Charged polyelectrolyte chain and counter-ions surrounded them

and chondroitin sulfate (polyanionic molecules) are frequently used [21, 29]. Also, neutral polysaccharides as dextran and cellulose are chemically modified to dextran sulfate and carboxymethylcellulose, respectively, to obtain ionic properties in the molecule [19, 29]. The desired characteristics of the vehicles obtained such as size, zeta potential, crosslinking and even morphology are related to the intrinsic ionic polysaccharide properties [21], which include the molecular weight, charge and distribution, and proportion of ionic groups. Temperature, pH, solvents, and mass ratio also need to be considered to obtain the expected drug loading and releasing over time, and in the specific site of ionically gelled systems [21, 22, 27]. The two most used methods for manufacturing ionically gelled particles for controlled drug release are described below.

2.1 *Ionic Crosslinking of Polysaccharides*

The ionotropic gelation or ionic gelation is a useful technique based on the combination of polyelectrolytes with counter ions to develop drug carrier devices [23, 31, 32]. This method is easy to use, works under mild conditions, and allows the use of biocompatible materials. Ionic gelation does not need toxic agents, as chemical cross-linkers, because the molecules interact only physically by electrostatic forces. Moreover, spherical structures in a wide range of sizes, from nano- to micrometers scale are generated [23, 31, 33, 34]. The devices obtained by this technique have been proved well performance to load and release different drugs and also have a large swell capacity, which facilitates the release [32]. To realize ionic gelation two solutions are needed: the first, the ionic polysaccharide solution dissolved in water or weak acid; the second, the counter ion solution. The latter should be in constant stirring, while the ionic polysaccharide solution is added by dropwise (the solutions can be inverted, and the counter ion solution would be dropping) [15, 23, 31–33]. Once the solutions are together, they are attracted by their opposite charges letting the ionic gelation by electrostatic interaction, while the stirring speed helps to determine the particle size. Spherical particles precipitate and are recovered by filtration and washed with distilled water or an ethanol/water mix [15, 31]. Figure 4 shows the steps for the ionotropic gelation method. The drug should be combined with the polymeric solution before the combination with the cross linker. Several factors must be considered for ionic gelation, for example, the ratio of ionic cross linker and the ionic polysaccharides, molecular weights, ionic strength and pH of the medium, temperature, and the drug and polymeric solution concentration [22, 32, 33].

The counter ions used as cross linker agents in ionotropic gelation are classified by the molecular weight, the valence number, or by the charge. According to their molecular weight, cross linkers are grouped in low molecular weight(e.g., $CaCl_2$, $BaCl_2$, $MgCl_2$, $CuCl_2$, $ZnCl_2$, $CoCl_2$, pyrophosphate, tripolyphosphate, tetrapolyphosphate, octapolyphosphate and hexametaphosphate) and in high molecular weight (e.g., octyl sulfate, lauryl sulfate, hexadecyl sulfate, cetylstearyl sulfate) [31]. Regarding to their valence number, they are divided into four classes: divalent (Ca^{2+}, Zn^{2+},

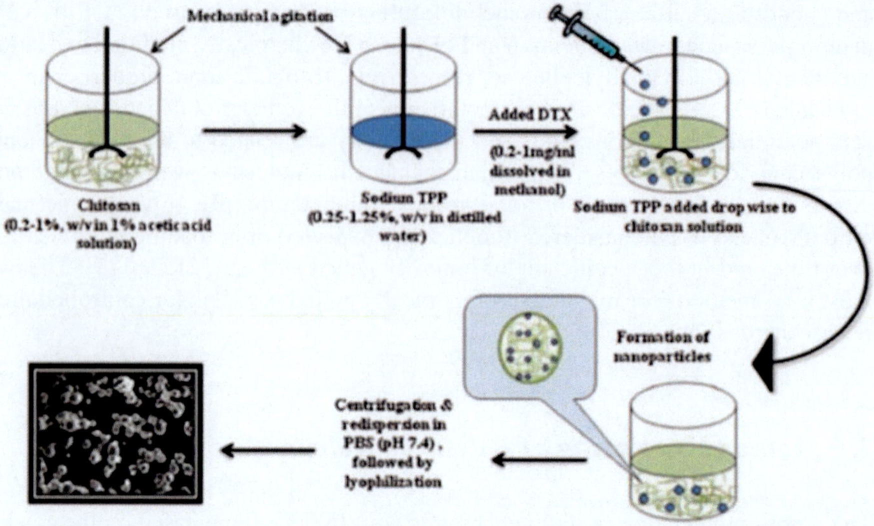

Fig. 4 Schematic representation of the ionotropic gelation method. *Source* Jain et al. [35]

and SO_4^{2-}), trivalent (citrate and PO_4^{3-}), tetravalent and pentavalent as ethylenediaminetetraacetic acid (EDTA), pyrophosphate (PPi), and tripolyphosphate (TPP) [33]. According to their charge, counter ions are divided into cationic cross linkers, including $CaCl_2$, $BaCl_2$, $MgCl_2$, $CuCl_2$, $ZnCl_2$, $CoCl_2$, and $AlCl_3$ or anionic cross linkers including pyrophosphate, tripolyphosphate, and tetrapolyphosphate [15]. The assembly between polyelectrolyte and the counter ion to form particles by electrostatic interaction is represented in Fig. 5.

2.2 *Polyelectrolyte Complexes Formation*

Other way to form ionic crosslinking particles from ionic polysaccharides is their combination with other opposite charged polyelectrolyte, leading to the polyelectrolyte complexes (PECs) formation (Fig. 6). In this case, the association and stability are due to the strong electrostatic interaction between the ionic groups distributed along the polyelectrolytes chains [22, 23, 27, 29, 36]. Besides, hydrogen bonding, van der Waals forces, and hydrophobic and dipole interaction are present in the complexation process [27, 29, 36]. According to researchers, the predominant driving force that leads to PECs formation is an increase in entropy level caused by the separation of counter ions and their release into the solution [22, 29]. PECs fusion as well as their physicochemical characteristics depend on the intrinsic properties of both ionic polysaccharides used (molecular weight and surface charge density) and on the preparation conditions (pH, temperature, polyelectrolytes ratio, and solvents nature) [22, 27, 29]. An advantage offered by this technique is the possibility of

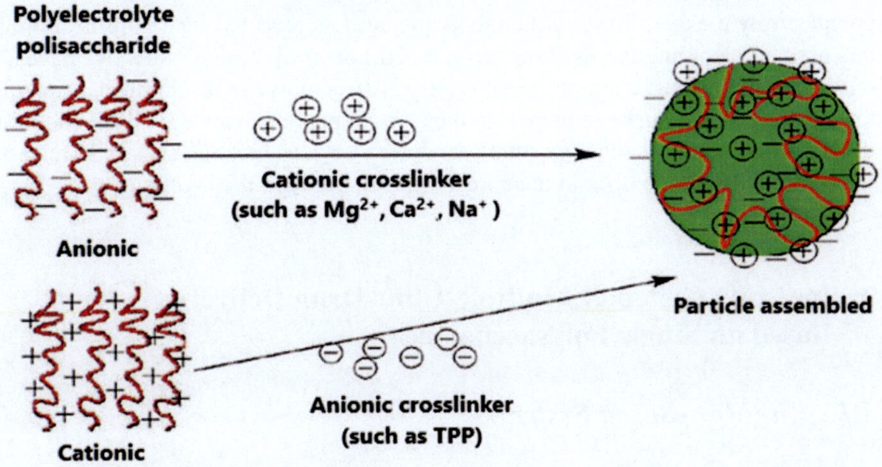

Fig. 5 Representation of ionic crosslinking of polyelectrolyte polysaccharides to form particles. *Source* Adapted from Debele et al. [22]

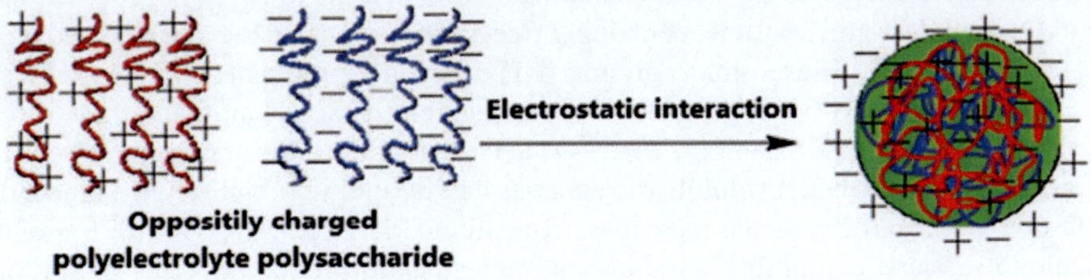

Fig. 6 Representation of PECs formation. *Source* Adapted from Debele et al. [22]

assembly colloidal or coacervate complexes when diluted solutions and low molecular weights are used [27, 29, 37]. The resulted PECs generally possess a combination of individual characteristics of each polymer. PECs are not neutral, they maintain a slight charge allowing to the particles to be dispersed in aqueous medium, helping to bind and interact with therapeutics molecules and cellular components [21, 36]. The surface charge results by an excess of functional groups charged, which are not involved in electrostatic interactions and are available to join with another oppositely charged polyelectrolyte [27, 30].

The process to assembly more than two polyelectrolytes (to do multilayers) is called the Layer by Layer method (LbL) and is a useful and straightforward way to produce several structures [29, 37]. Initially, this technique was employed in the thin film production based on the alternating deposition of opposite polyelectrolytes onto a charged surface, leading to the conversion of surface charge [29]. Today, LbL has been adapted to cover micro and nanosized particles, starting with colloidal structures and then adding polyelectrolyte solutions to cover it. The particles are recovered by centrifugation, and several washing steps are required to remove the polyelectrolyte excess [38]. Although, some authors reported the simplification of this method, avoiding the washing steps repetitions [38]. Among the advantages, a relatively easy construction of a great variety of structures can be obtained, changing the numbers of layers, the sequence, or the types of polyelectrolytes [29]. In addition, the LbL process offers the opportunity to design devices for drug sustained-release, which encapsulate the drug in the core as well as between the layers.

3 Ionically Gelled of Multiple-Units Drug Delivery Systems Based on Single Polysaccharide

3.1 Chitosan-Based Systems

The chitosan is the only cationic polysaccharide in nature, considered one of the most beneficial biopolymers for drug delivery systems designing [23, 29, 39]. It is obtained from crustaceous shells after the deacetylation process of chitin. Their chemical structure is formed by linear chains with random units of β-1,4-D-glucosamine and N-acetyl-D-glucosamine, containing free amine groups on the deacetylated unit [19, 29, 31, 40]. These amino groups (NH_2) become protonated (NH_3^+) in acid solution (below pH 6), as a function of the degree of deacetylation and the molecular weight of the chitosan [23, 29, 39–41]. The cationic chitosan presents essential properties such as better solubility in aqueous media, antimicrobial activity, mucoadhesion and can increase the molecule permeation [19, 21, 23, 25, 39, 40]. Cationic chitosan is also combining with anionic molecules to produce devices of several forms and dimensions. Mainly, the researchers are focusing on nano- and microparticles, which serve as a vehicle for drug transportation inside the body [25, 31]. The most common combination is chitosan with sodium tripolyphosphate (TPP), through ionotropic gelation. However, several works are using different cross linker agents as citrate, pyrophosphate (PPi), calcium phosphate $Ca_3(PO_4)_2$, sodium sulfate, and β-glycerophosphate [19, 33, 36]. All of them have the capacity to be joined to the amino groups by electrostatic interaction when they are between the chitosan backbones to form a physically cross-linking structure.

At this moment, many kinds of research focused on the production of ionically gelled chitosan-based systems for drug delivery have been carrying out due to the significant advantages offered by both the polymer and the method. Different kinds of active compounds such as hydrophobic, anticancer, or colon target drugs, as well as peptides and proteins, have been encapsulated in these systems for protecting, targeting, and releasing them in a sustained manner. The chitosan-based systems vary in structural dimensions from nanoparticles, microspheres to beads. Generally, this kind of structure has shown to be pH-sensitive for swelling (high swell in acidic medium and shrunk in neutral or alkaline medium) or for drug releasing at a specific place [42]. The ionotropic gelation technique has been combined with spray drying, inverse emulsion, high voltage electrostatic field, even with chemical crosslinking to produce chitosan-based systems [42–44]. According to some studies, parameters as chitosan molecular weight, mass ratio of chitosan:ionic cross linker agent, and the salinity of initial solutions are the main factors that determine the final characteristics of chitosan-based particles [35, 45, 46]. Some characteristics of particles obtained by ionotropic gelation, based on chitosan-TPP combination are describing below.

3.1.1 Morphology, Size and Zeta Potential of Ionically Gelled Chitosan-TPP Based Particles

Mostly of chitosan-based particles produced by ionotropic gelation using the TPP as cross linker agent have shown a well-defined spherical shape by light microscopy, SEM or TEM. Chitosan-TPP nanoparticles generally have presented diameters from 150 to 350 nm, although others chitosan-TPP nanoparticles have had diameters between 80 and 100 nm [35, 47–51]. The size of these type of nanoparticles can be affected by the concentrations of the initial solutions. According to Jain et al.] [35], the increase in the concentration of the chitosan and the drug solutions caused the formation of larger nanoparticles, while the increase in TPP concentration caused more compact and smaller structures. Dustgani et al. [48] also found that nanoparticle size increase with the increase of drug concentration. Nguyen et al. [44] reported that the use of the high molecular weight of chitosan increased the size of chitosan-TPP nanoparticles by more than 1000 nm and presented a more significant variation in size. Some authors reported that the size and polydispersity index of the chitosan-TPP nanoparticle could be reduced by the addition of salts, such as NaCl or NaF to the initial solutions (Fig. 7) [45, 52]. The zeta potential of ionically gelled chitosan-TPP nanoparticles frequently have been reported between 30 and 50 mV, indicating a positive charge at the surface due to the amino groups that remain free after neutralization with the anionic molecule. Sometimes zeta potential value can change due to the presence of charged drug molecules, as happened in the work of Hassani et al. [49] when incorporating the tacrine drug. The presence of the surface charge is essential to maintain the stability of nanoparticles in solution due to it cause an electrostatic repulsion [35].

Regarding the size of chitosan-TPP microspheres obtained by ionic gelation, they have displayed a wide range, mainly caused by the variation of the formulation conditions, the chemical characteristics of the polysaccharide, and by the active component enclosed. Ma et al. [43], produced chitosan-TPP microspheres by ionotropic gelation under high voltage electrostatic field. The particles presented diameters from 100 to 450 μm depending on the concentration of the chitosan solution that ranged between 0.5 and 1.2 g/L. The surface morphology was rough, and the authors associate it with the addition of ethanol to TPP solution to obtain a complete coagulation (Fig. 8). Similar morphology was observed by Zou et al. [42] in chitosan microspheres with sizes around 5–10 μm made by inverse emulsion-ionic crosslinking with TPP molecule. However, after loading the BSA as a drug model, microspheres showed a smooth surface (Fig. 9). Contrary, other empty chitosan-TPP microspheres presented a smooth surface, while those loaded with resveratrol were rough. These

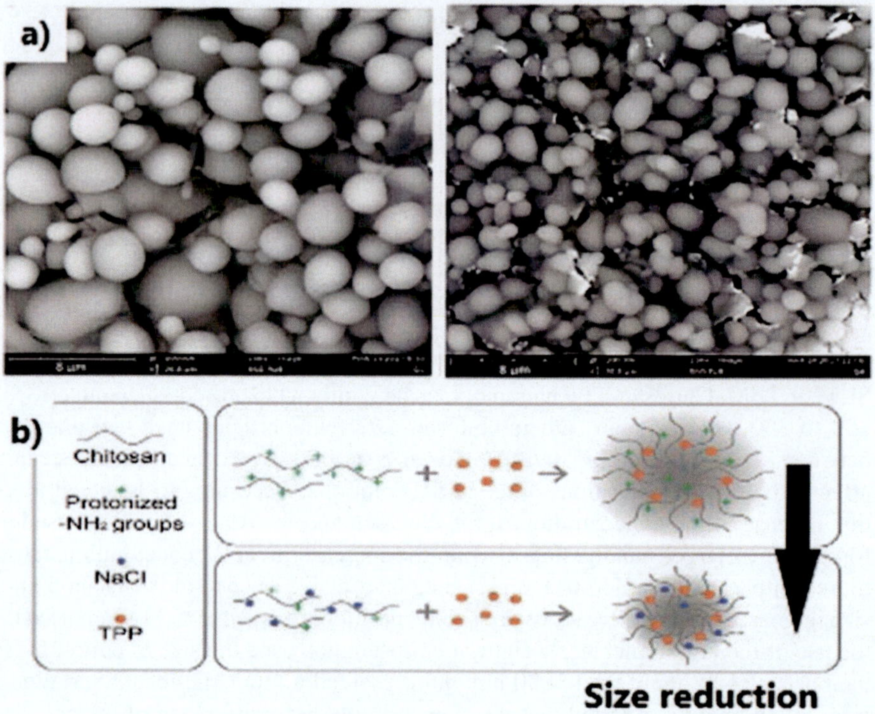

Fig. 7 Effect of adding salt on particle size during the ionotropic gelation procedure procedure **a** SEM images are showing the size change of chitosan-TPP nanoparticles. **b** Graphic representation of the effect at molecular level. *Sources* Adapted from Antoniou et al. [45] and Farias da Silva Furtado et al. [52]

Fig. 8 SEM micrograph of chitosan-TPP microparticles obtained by ionic gelation in ethanol medium. *Source* Adapted from Ma et al. [43]

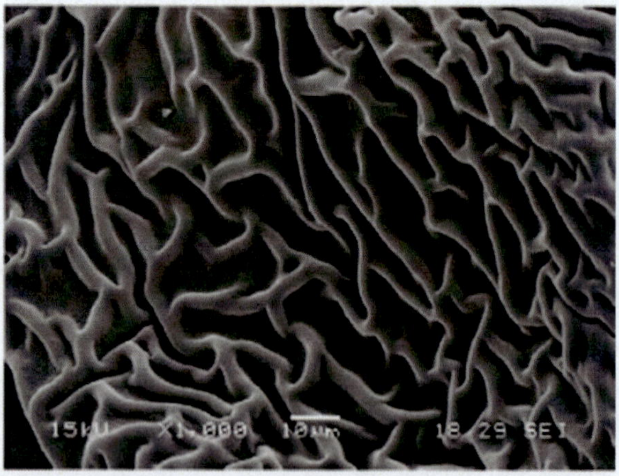

Fig. 9 SEM micrograph of chitosan-TPP microspheres. **a** Chitosan-TPP microspheres and **b** BSA-loaded chitosan-TPP microspheres. *Source* Adapted from Zou et al. [42]

microspheres had sizes from 160 to 200 μm, and less variation in sizes was observing when the TPP solution concentration increased [53]. Chitosan-TPP beads made by ionic gelation showed a size around 4–1.5 mm. However, the lower the polymer concentration, the smaller the beads size and it remained more or less constant at higher TPP concentrations [54].

3.1.2 Drug Loading and Releasing of Ionically Gelled Chitosan-TPP Based Particles

Chitosan-TPP ionically gelled particles are widely studied as drug delivery systems to treat several diseases as diabetes, cancer, tuberculosis, hypertension, and microbial infections. Among the active substances loaded into chitosan systems are insulin, amoxicillin, tacrine, docetaxel, letrozole, bovine serum albumin (BSA), corticosteroid, anti-hypertensive biopeptides, resveratrol, and scorpion venom. The amoxicillin antibiotic was loaded into chitosan-TPP nanoparticles with an encapsulation efficiency (EE) of 90% and presented better antibacterial activity than the alone amoxicillin. Notably, small size nanoparticles and with high zeta potential were better [44]. The great bactericide effect against *Streptococcus pneumonia* of the chitosan-TPP amoxicillin-loaded nanoparticles allowed to reduce until three times the antibiotic dosage. On the other hand, an anti-hypertensive biopeptide was loaded into chitosan-TPP nanoparticles, and their therapeutic effect was tested in hypertensive rats. The EE of these systems was 75%. A decreased particle size (160 nm) caused by the speed and agitation time led to better EE. The in vitro sustained release of this biopeptide in pH 4 medium was 58% after 12 h of incubation. The in vivo studies proved a blood-pressure-lowering effect dosage-dependent. The chitosan-TPP nanoparticles improved the stability and bioactivity of the anti-hypertensive biopeptide, showing a high anty-hypertensive effect compared with the un-encapsulated peptides [47].

The anticancer drugs, letrozole and docetaxel were encapsulated with high efficiency in ionically gelled chitosan-TPP nanoparticles. For letrozole, the EE was upper than 90% and for docetaxel was 78%. The release percentage for both drugs was 78% in buffer phosphate at pH 7.4 and 37 °C (Fig. 10) [35, 55]. The docetaxel-loaded nanoparticles showed a significant reduction of cancer-cells viability (85%), decreasing viability by 25% more than non-encapsulated docetaxel [35]. Dexamethasone sodium phosphate, a drug used to treat aphtha lesions, have been incorporated into chitosan-TPP nanoparticles with an EE up to 76% [48]. These nanoparticles also are suitable drug systems to carry tacrine (anti-Alzheimer's drug), showing an EE of 66% [49]. Insulin loading was also successful with EE of 73% and their releasing in pH 7.4 was 60% [50]. Patel et al. [51] developed an oral sustained release chitosan-TPP nanoparticles for rifampicin, which encapsulated until 35% of the drug and release around 85%. Another chitosan-TPP nanoparticles showed to have a high EE (91.9%) to load scorpion venom (*Mesobuthus eupeus*) and release it until 60% along 10 h. The most common release mechanism that primarily govern these systems are the Fick diffusion combined with the polymer dissolution, and the polymer swelling [35, 50, 51].

The ionically gelled chitosan-TPP microspheres have been studied as a drug carrier system for resveratrol. They showed a high EE of 94.5%, mainly when a low chitosan concentration (0.5%) was used. The burst release effect of resveratrol decreased by the increased in TPP solution concentration and a cumulative release around 80% was reached at 7 h [53]. The pH-sensitive release behavior of BSA-loaded chitosan-TPP microspheres was evaluated at different pH. Buffer solutions at pH 1, 3, and 5 were used and the released BSA values obtained after 15 days were 99%, 80%, and 38%, respectively. A low percentage (19%) was released when BSA-loaded microspheres were in pH 7, while 27% of BSA was released in pH 9 (Fig. 11) [42]. Ma et al. [43] also reported an EE of BSA superior to 90% in these types of

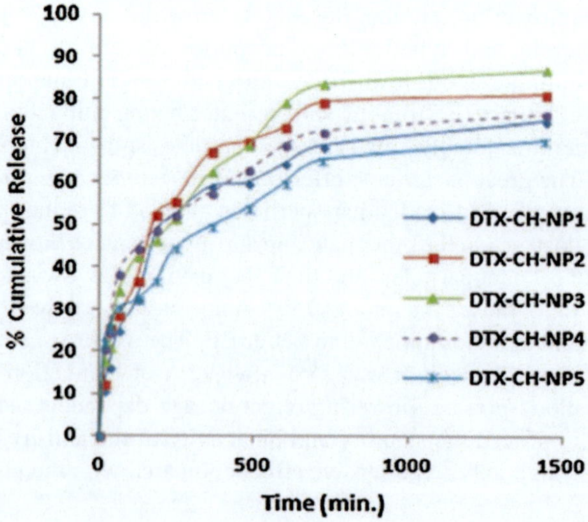

Fig. 10 Release profile of docetaxel (loaded at different concentrations) from chitosan-TPP nanoparticles. *Source* Jain et al. [35]

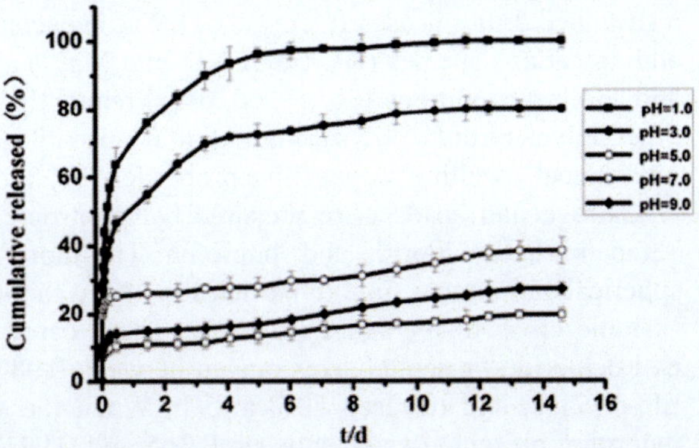

Fig. 11 Cumulative release of BSA from chitosan-TPP microspheres at different pH. *Source* Zou et al. [42]

systems when BSA: chitosan weight ratio was below 5%. However, the maximum level of BSA released was 50% at pH 8.8 and only 15% of BSA was released at pH 4. Despite the hopeful expectations of the chitosan-TPP particles produced by ionic (ionotropic) gelation, their final application has not been achieved due to the lack of reproducibility inter e intra laboratories. Many times the particle preparation is made with different batches of chitosan, which can vary in physical properties and biological responses [41, 46]. Some authors recommended to carry an extensive characterization of chitosan batch before using and better experimental design [29, 41].

3.2 Carrageenan-Based Systems

Carrageenan is a group of natural sulfated polysaccharides extracted from red seaweeds (Rhodophyceae class) which possess attractive structural and functional properties for pharmaceutical and biomedical areas [19, 56–59]. Structurally, they are composed by randomly oriented linear chains of β-D-galactose and 3,6-anhydrous-α-D-galactose joined by α-1,3 and β-1,4-glycosidic bonds, which contain sulfated groups. The sulfated groups vary in quantity (15–40%) and position among carrageenan and they are responsible for the anionic character of these polymers [19, 56–59]. Depending on the seaweeds species used for carrageenan isolation and the sulfate proportion in the chain, they are classified into six kinds: Kappa (k), Iota (ι), Lambda (λ), Mu (μ), Nu (ν), and Theta (θ) being the first three the most important commercially. Especially, k and ι are prety commercial because their good gelling properties [20, 56–58, 60]. Usually, carrageenans have been used as stabilizer in food industry, but recently several biological activities as immunomodulatory, antitumor, anticoagulant, antioxidant, antiviral, and anti-hyperlipidemic have been related to them [20, 56–58, 60]. Moreover, carrageenans present notably physicochemical properties that enabling them to form useful vehicles for drug delivery with the pH- and temperature-sensitiveness characteristics. These vehicles proved to have a sustained drug release effect [19, 61]. Carrageenan ionotropic gels are formed by adding cations species (K^+, Ca_2^+, Na^+ and Mg_2^+) as physical cross linkers agents and cooling conditions [56, 57, 60, 61]. Frequently, carrageenan is combined with other polyelectrolyte polysaccharides to form stable particles, which can have better mechanical, swelling, or sensitive properties [19, 56, 60].

Carrageenan beads were prepared by ionotropic gelation for encapsulation of verapamil-hydrochloride and ibuprofen. The more suitable formulation to create spherical and almost smooth particles included the k-carrageenan and the KCl salt as ionic cross linker agent (Fig. 12). The k-carrageenan beads with verapamil-hydrochloride presented sizes ranged between 0.980 and 1.197 mm, displayed an EE of 71%, and released 70% at 5 h. When the same beads were loaded with ibuprofen presented sizes between 0.865 and 0.921 mm, an EE of 58% but only released 30% at 6 h (Fig. 13) [62]. Similar carrageenan-KCl microspheres systems were assembled to encapsulate the ciprofloxacin HCl antibiotic. The carrageenan-KCl microspheres exhibited spherical shape with a size of 1.67 μm, had a smooth surface, and were able to encapsulate only 29% of ciprofloxacin HCl. However, the release studies were not conducted [63]. In other studies, active compounds have also been encapsulated in carrageenan particles to be preserved and delivered, such as carotenoids, which are associating with numerous health benefits. Soukoulis et al. [64], prepared k-carrageenan particles using three different ions (Na^+, K^+, or Ca^{2+}) to crosslinking the systems. However, the K^+ ion generated stronger particles than the other two. Besides, K^+ ion and Na^+ ion, exerted the highest retention and better bioaccessibility of β-carotene. Carboxymethylated k-carrageenan has been used to produce insulin-loaded microparticles. Also, they were functionalized with lectin to enhanced intestinal mucoadhesiveness. They presented a high EE of insulin with 94%. The in vitro release study exhibited that only 4.2% of insulin was released at acidic conditions, but in pH 6.8 the complete insulin was released in 10 h. The

Fig. 12 SEM micrograph of loaded k-carrageenan beads: **a** with verapamil hydrochloride and **b** with ibuprofen. *Source* Adapted from Sipahigil and Dortunc [62]

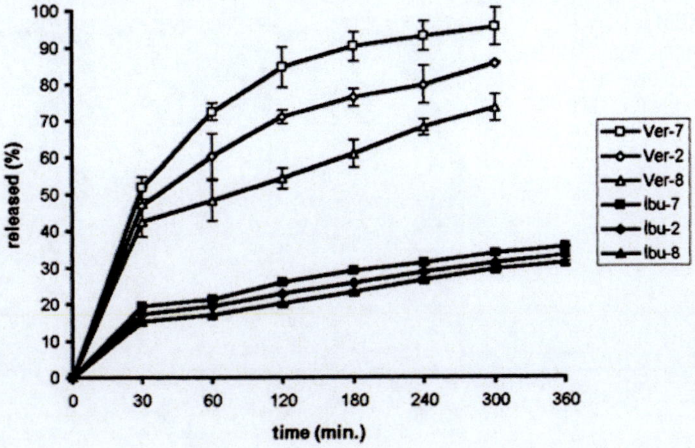

Fig. 13 Release profiles of verapamil and ibuprofen from carrageenan beads prepared at different KCl concentrations. *Source* Sipahigil and Dortunc [62]

hypoglycemic effect of insulin-loaded microparticles was proved after their oral administration in diabetic rats during 12–24 h [65].

3.3 Dextran Sulfate-Based Systems

Naturally, dextran is a water-soluble and neutral polysaccharide produced by bacteria (*Leuconostoc mesenteroides* and *Streptococcus mutans*) from a sucrose source and it has medical applications. However, some structural modifications have been made on it to develop the polyelectrolyte character [21, 66]. Dextran is biocompatible and biodegradable to the human body because its structure only consists of glucose linked by α-1, 6 glycosidic bonds, and presents branches joined by α-1,3 bonds [21, 23, 29]. Dextran molecule acquires anionic character by adding sulfate groups to the chain (around two by glucose unit), becoming a dextran-sulfate molecule [21, 29]. This polysaccharide is sold as dextran sulfate of sodium and performs similar anticoagulant activity to the heparin but at lower cost [29]. Intensely, few studies have focused on the development of particles based on the ionic cross-linking of this polymer. In 2015, Kutscher et al. [67], formed nanoparticles based on electrostatic interactions between the polyanionic dextran sulfate and one cationic small drug molecule, ciprofloxacin, and evaluated the effect of salt incorporation into the mix. The results showed that the addition of NaCl caused huge quantities of spherical and small particles (383 19 nm) with smooth surface and negative charge of −50 mV (Fig. 14). The results indicated that the NaCl incorporation led to a high binding affinity of the drug to the polyelectrolyte to form nanoparticles, and generated a high releasing of ciprofloxacin at only 15 min. On the other hand, PECs were formed

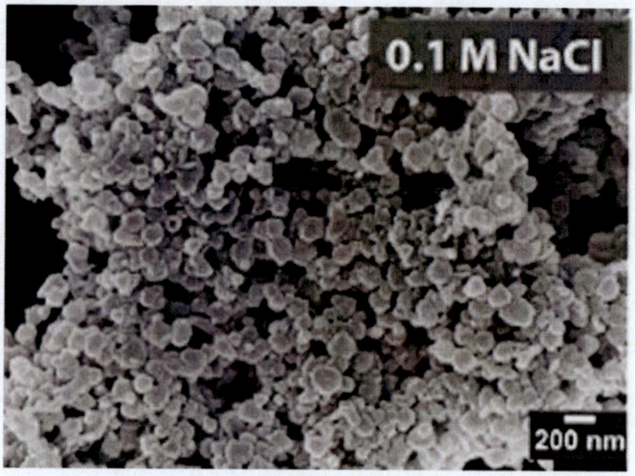

Fig. 14 SEM images of dextran-sulfate nanoparticles obtained by ionotropic gelation in the presence of salt (NaCl). *Source* Adapted from Kutscher et al. [67]

from dextran-sulfate and chitosan to assembly beads, micro- or nanoparticles as drug carrier systems, having the peculiarity to be easy metabolized by liver or spleen [23, 29, 66].

4 Multiple-Units Drug Delivery Systems Based on Ionic Gelation of Several Polymers

The production of PECs-based drug release systems has become one of the most recurrent alternatives in the research area due to the properties offered by several ionic polymers and the easy way to combine them through the ionic gelation technique. Most of the polysaccharides-based PECs used chitosan or their products (modifications or oligomers) to combine with almost all polyanions due to the advantageous properties and their polycationic character [21, 36]. In general, the combination of two polyelectrolytes allows neutralizing the ionic groups, expecting to get better properties of the new materials [68]. The complexation between ionic polysaccharides affects morphology, size, swelling capacity, and functional characteristics [15, 68]. The more variations in the polymer combinations, the greater variety in the PECs characteristics. Among the principal polysaccharides mixed with chitosan are hyaluronic acid, alginate, dextran sulfate, carrageenan, chondroitin sulfate, pectin, xanthan gum, collagen, heparin, and carboxymethyl cellulose [23, 27, 29, 36].

The assembly of chitosan-dextran sulfate has resulted in one of the most popular PECs to produce particles with potential as drug carriers. Lin et al. [69] reported the assembly of microspheres composed by chitosan, TPP, and dextran sulfate to transport ibuprofen to the colon. The particles had around 870-950 μm in diameter, high porosity surface structure (Fig. 15), and large swelling capacity (80%), both caused by the increase of dextran sulfate content. The EE was higher than 80%, while the release profile was slow at pH 1.4 and very fast at pH 6.8, releasing almost 100%

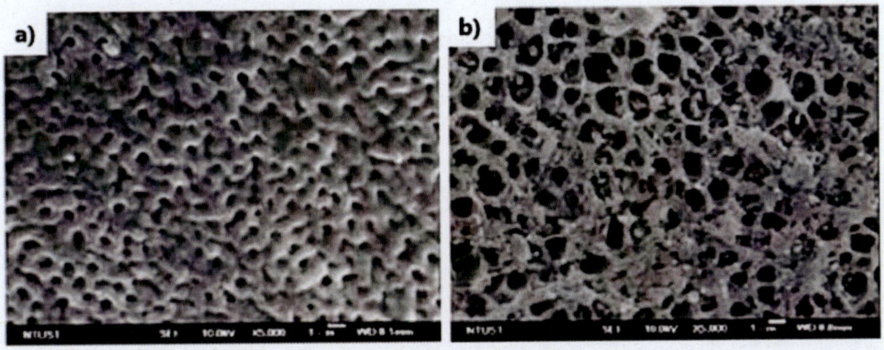

Fig. 15 SEM micrographs of chitosan microspheres surfaces: **a** only chitosan-TPP and **b** PEC of chitosan-dextran sulfate and TPP. *Source* Adapted from Lin et al. [69]

of loaded ibuprofen. On the other hand, mucoadhesive nanoparticles of chitosan-dextran loaded with insulin were prepared and analyzed by changing the polymer ratio. At the ratio of 1:9 chitosan-dextran sulfate, the nanoparticles were 432 nm in diameter and present zeta potential of -54 mV, while at the ratio of 6:4 the size was 320 nm and the zeta potential +38.6 mV. They showed low EE for insulin encapsulating only 9.73% and 16.7% in the 1:9 and 6:4 chitosan/dextran sulfate ratios, respectively. However, they release around 80% of loaded insulin at 32 °C in a Franz diffusion cell system [70]. Other authors formed chitosan-dextran sulfate nanoparticles with ciprofloxacin as an ophthalmic application. The loaded particles were spherical and presented a size of 350 nm and zeta potential of +3.55 mV. The nanoparticles were able to encapsulate 83% of the ciprofloxacin and released around 90% after 24 h. Indeed, they were tested against Gram-positive and Gram-negative bacterium, and ciprofloxacin-loaded nanoparticles appeared more effective than the alone drug [71].

Another popular combination to form PECs is the mix between chitosan and alginate. Microparticles of chitosan-alginate PECs containing the antibiotic vancomycin chloride were creating by Unagolla et al. [72]. The microparticles were spherical with a diameter of 680–690 μm, presented a rough surface and their EE was only about 11–13% (Fig. 16). When microparticles were lyophilized, they had more swelling capacity and showed the best-controlled release of vancomycin with a releasing average of 22 μg per day for 14 days. The drug transport mechanism was controlled by both Fickian diffusion and case II relaxations. In other study, nanoparticles based on these two polyelectrolytes were fabricated to enhance the antidiabetic activity of curcumin when it was encapsulated. The nanoparticles size was < 50 nm. The chitosan-alginate nanoparticles had a porous structure and presented an EE of curcumin of 76%, which was higher than those nanoparticles formed by chitosan-TPP. The chitosan-alginate PECs reduced the loss of curcumin by 20% and showed a significant reduction of hyperglycemia within 7 days of treatment [73]. A preparation of dual ionic crosslinked alginate-chitosan beads with calcium chloride and sodium sulfate was made to load BSA. The beads showed an irregular morphology with a

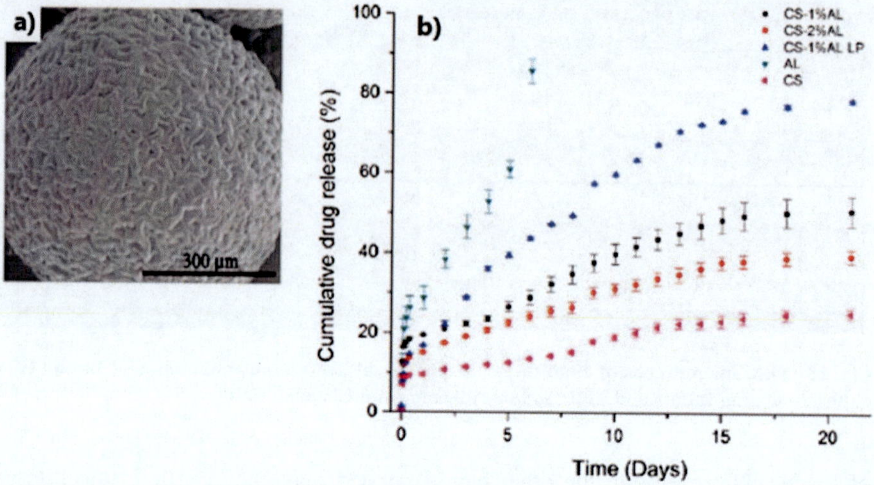

Fig. 16 Chitosan-alginate PEC microparticles. **a** SEM micrograph of the surface and **b** Cumulative release profile of vancomycin from the microparticles. *Source* Adapted from Unagolla et al. [72]

porous surface as chitosan increased, the average size was around 1.5 mm and it was higher with the increase of chitosan. They also presented higher swelling when pH or alginate content increased. Regarding the release behavior, the dual crosslinked beads released around 2%, 80%, and 97% at pH 1.0, 6.8, and 7.4, respectively, suggesting that the dual crosslinked beads presented a potential system at small intestine or colon site-specific drug delivery [74].

An alternative way to produce PECs is making modifications to the structure of polymers. Curdlan, a neutral and linear (1-3)-β-glucan from *Agrobaterium* [75], has been modified by Yan et al. [76] to perform a curdlan sulfate with a negative charge and subsequently to form a curdlan sulfate-chitosan PEC for releasing the antiviral zidovudine. The obtained nanosized PEC showed a spherical morphology with a smooth surface, an average diameter of 180 nm and zeta potential of −38 mV. The EE of zidovudine into the nanoparticles was 80% and their release profile was pH-dependent, since at pH 7. 4 only 38% was released, and at pH 4.5 was released up to 72%. Another commonly used polymeric modification to provide anionic character is the carboxymethylation. In this case, carboxymethyl-starch was synthesized to combine with chitosan and to form a PEC vehicle of BSA, used as a model drug. The results were oval microparticles of 1.53 mm with rough and porous surfaces, even more than those prepared with chitosan-TPP (Fig. 17a, b). These microspheres presented an EE of BSA of 67%, whereas their releasing was 58% at pH 1.2 (simulating gastric conditions) and only 15% at pH 6.8 (simulating intestinal conditions), both during 12 h. The comparison between carboxymethyl-starch and chitosan with those of chitosan and TPP showed that the release profile was more controlled by the PEC of carboxymethyl-starch and chitosan (Fig. 17c, d). Moreover, at pH 6.8, the release followed the zero-order kinetics becoming this kind

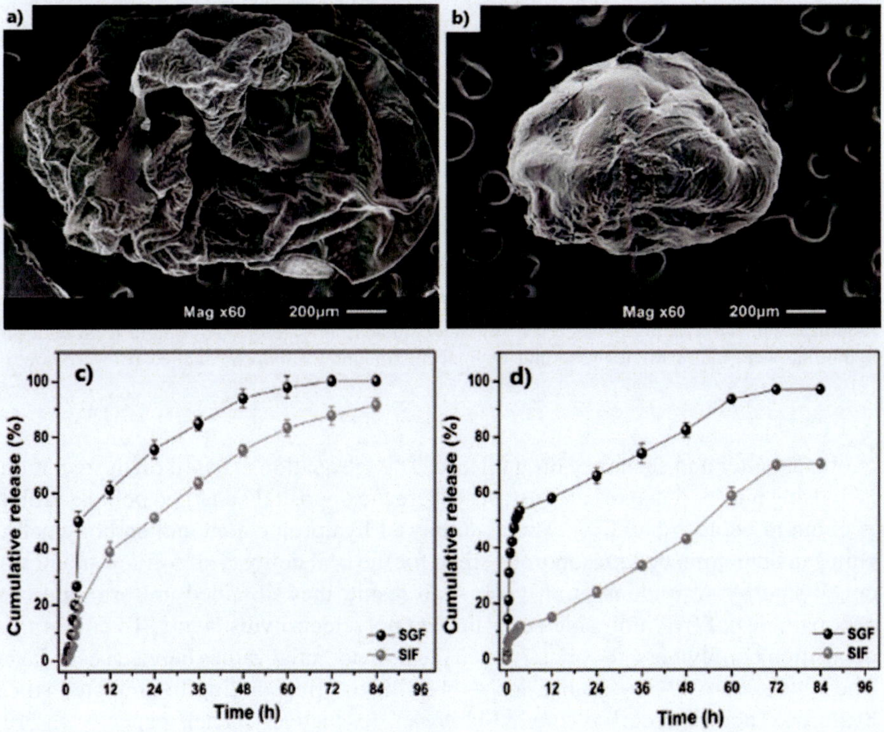

Fig. 17 Above: SEM micrographs of microparticles based on carboxymethyl starch-chitosan (PEC) (**a**), and in chitosan-TPP (**b**). Below: The release profiles of BSA from microparticles based on the PEC (**c**) and in the chitosan-TPP microparticles (**d**). *Source* Adapted from Quadrado and Fajardo [68]

of PEC an attractive system for drug delivery at specific regions of the intestinal tract [68]. In other cases, chitosan has been hydrophobically modified with deoxycholic acid to be a hydrophobic-drug carrier combined with TPP and hyaluronic acid. The results were spherical nanoparticles with sizes around 280–310 nm, they decreased in size with the increase of the chitosan degree substitution (DS), as well as the zeta potential vary by 30.7–18.2 mV. However, the contrary effect was observed when doxorubicin was loaded and released because as DS increased the EE was higher (around 38–57%), and the drug release increased about 35–50% without burst effect (Fig. 18). They showed biocompatibility even at high concentrations [77].

4.1 Multilayered Polyelectrolyte Complexes Assembly

Taking advantage of the characteristics and the easy way to perform the PECs, some researchers have gone further and alternated layers of polysaccharide polyelectrolyte

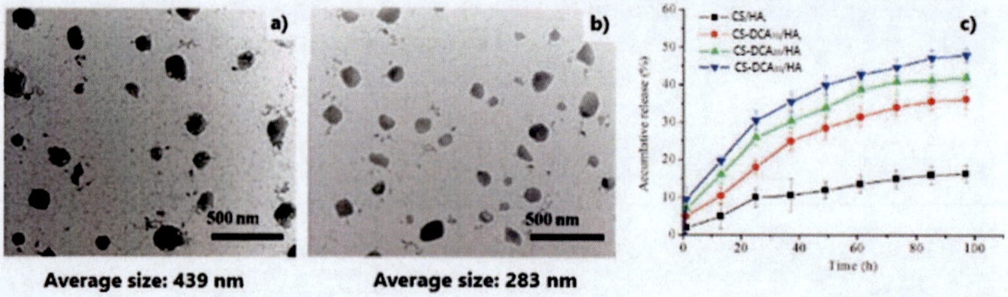

Fig. 18 TEM micrographs of nanoparticles based on chitosan-hyaluronic acid PEC (**a**) and modified chitosan-hyaluronic acid PEC (**b**). Accumulative release of doxorubicin from chitosan-hyaluronic acid nanoparticles with and without modification. *Source* Adapted from Chen et al. [8]

to obtain better and sustained drug release. The fabrication is based on the technique LbL using micro- or nanoscale surfaces where the oppositely charged polyelectrolyte is going to be adsorbed [78]. The assembly of hyaluronic acid and carboxymethyl chitosan onto aminated mesoporous silica for the oral delivery of 5-fluorouracil was developing by Anirudhan et al. [79]. As a result, they obtained uniform and oval nanoparticles of 290 nm, which contain two polyelectrolytes layers. To confirm the deposition of polymers, they determined the zeta potential values between each layer. The values were +33, −32.68, and +44.54 mV corresponding to aminated silica, hyaluronic acid and carboxymethyl-chitosan, respectively. Their capacity as drug delivery system was proved because they presented an EE of 78% at pH 5.4 and a cumulative release around 80% at pH 7.4. Indeed, they had antitumor activity. On the other hand, Zhang et al. [80] produced nanocapsules through electrostatic LbL method alternating the anionic carboxymethyl starch and the cationic quaternary ammonium starch onto BSA particles used as support, as well as the active core. The final nanoparticles were assembled with three polyelectrolyte layers, which were affected by the pH changes, as observed in Fig. 19. The use of polymers with low substitution degree and low molecular weight produced a more compact and stable

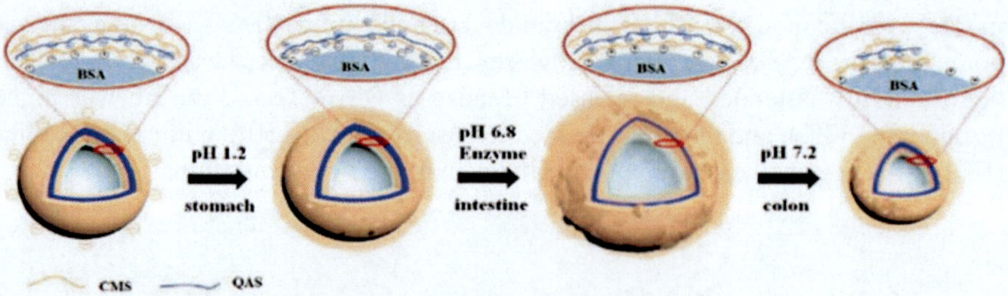

Fig. 19 Schematic representation of nanoparticles assembled by layers of modified starches exposed at different pH. *Source* Zhang et al. [80]

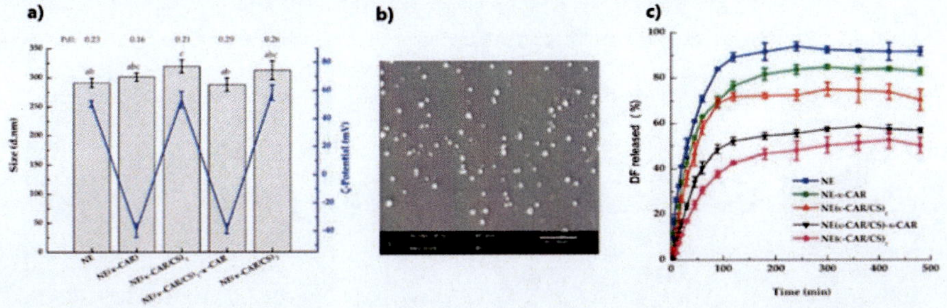

Fig. 20 Characteristics of chitosan-kappa carrageenan nanoparticles. **a** Sizes and zeta potentials after the adsorption of polymer layers. **b** SEM micrograph. **c** Release profiles of diflunisal from the different nanocapsules. *Source* Adapted from Rochín-Wong et al. [81]

core-shell structure. Also, they could exhibit better colon-specific delivery due to at pH 7.2 the 65% of BSA was sustainedly released.

In other work, layers of k-carrageenan and chitosan (bilayer) were alternately deposited onto olive oil nanoemulsion to assembly a release system of the anti-inflammatory diflunisal drug. After four layers, the nanocapsules size was 300 nm, with spherical morphology. The adsorption of each polysaccharide layer was evidenced by charge reversal from −30 to +45 mV (Fig. 20a, b). The EE of diflunisal on these nanocapsules were of 72%, and their release profile is determined by the number of polyelectrolyte layers. The 80% of diflunisal was released at 60 min when only a bilayer was adding, while the same percentage was released at 120 min when two bilayers were deposited (Fig. 20c) [81]. In another case, Jeon et al. [82], multi-layered liposomes were engineered with up to 10 alternating layers of hyaluronic acid and chitosan, using dihexadecyl phosphate as a template with a negative charge (Fig. 21). The final liposomes size was 520 nm. They presented spherical form and changed their zeta potential as polyelectrolyte changed. They were using as a carrier for the antioxidant quercetin, which was better released at pH 5 and in a sustained manner as more polymer layers were added. Besides, these nanoliposomes showed to improve the skin permeability of quercetin, displaying potential as transdermal drug delivery system.

As a conclusion, the ionotropic gelation method and the polyelectrolyte complex formation are suitable forms to produce multiple-unit systems for oral drug administration as well as for others routes. In both cases, the preparation is easy and under mild conditions, allowing the preservation of individual properties of each component. Indeed, the fabrication of these systems with ionic polysaccharides makes them appropriated for their usage in biomedical, pharmaceutical, and cosmetic areas. Through these procedures, biocompatible, no toxic, and biodegradable particles as drug delivery systems can be constructed in wide range of size and physicochemical characteristics. Better than that, these techniques help to obtain devices to protect the drug and facilitate its arrival in a target site, where they can perform a sustained

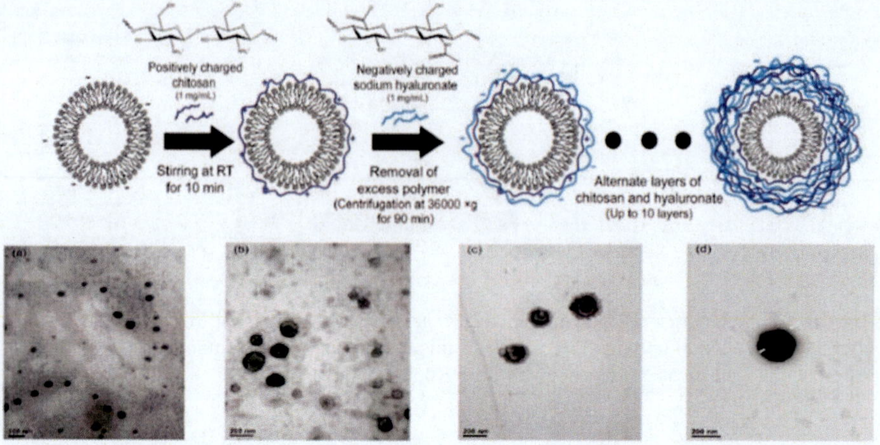

Fig. 21 Schematic representation of the fabrication of nanoparticles and TEM images of nanoparticles in each step. Source Adapted from Jeon et al. [82]

release. However, it is necessary to homogenize and describe adequately the fabrication and characterization protocols in order to reduce the variability of the generated particles, so they can be replicated at industrial level and have an adequate implementation.

References

1. Gupta H, Bhandari D, Sharma A (2009) Recent trends in oral drug delivery: a review. Recent Pat Drug Deliv Formul 3(2):162–173. https://doi.org/10.2174/187221109788452267
2. Nayak AK, Hasnain MS (2019) Background: multiple units in oral drug delivery. In: Plant polysaccharides-based multiple-unit systems for oral drug delivery. SpringerBriefs in Applied Sciences and Technology. Springer, Singapore, pp 1–17. https://doi.org/10.1007/978-981-10-6784-6
3. Debotton N, Dahan A (2017) Applications of polymers as pharmaceutical excipients in solid oral dosage forms. Med Res Rev 37(1):52–97. https://doi.org/10.1002/med.21403
4. Hamman H, Hamman J, Steenekamp J (2017) Multiple-unit pellet systems (MUPS): production and applications as advanced drug delivery systems. Drug Deliv Lett 7(3):201–210. https://doi.org/10.2174/2210303107666170927161351
5. Reinholz J, Landfester K, Mailänder V (2018) The challenges of oral drug delivery via nanocarriers. Drug Deliv 25(1):1694–1705. https://doi.org/10.1080/10717544.2018.1501119
6. Homayun B, Lin X, Choi HJ (2019) Challenges and recent progress in oral drug delivery systems for biopharmaceuticals. Pharmaceutics 11(3)s. https://doi.org/10.3390/pharmaceutics11030129
7. Padhee K, Chowdhary KA, Pattnaik S, Sahoo SK, Pathak N (2011) Design and development of multiple-unit, extended release drug delivery system of Verapamil HCL by pelletization technique. Int J Drug Dev Res 3(3):118–125
8. Chen T, Li J, Chen T, Sun CC, Zheng Y (2017) Tablets of multi-unit pellet system for controlled drug delivery. J Control Release 262:222–231. https://doi.org/10.1016/j.jconrel.2017.07.043

9. Efentakis M, Koutlis A, Vlachou M (2000) Development and evaluation of oral multiple-unit and single-unit hydrophilic controlled-release systems. AAPS PharmSciTech 1(4)
10. Uhumwangho M, Okor R (2008) Estimation of the Release profiles of multi-unit dose tablets of theophylline from the release profiles of their components. Trop J Pharm Res 7(2):981–986. https://doi.org/10.4314/tjpr.v7i2.14682
11. Ramu S, Ramakrishna G, Balaji M, Kondala K, Haranadh S (2013) Multiple unit drug delivery system: pelletization techniques. Am J Adv Drug Deliv 1(1):11–21
12. Al-Hashimi N, Begg N, Alany RG, Hassanin H, Elshaer A (2018) Oral modified release multiple-unit particulate systems: compressed pellets, microparticles and nanoparticles. Pharmaceutics 10(4). https://doi.org/10.3390/pharmaceutics10040176
13. Heimy Franceline Martínez Sánchez, AYEL, Méndez-Gómez, E, Vázquez AE, de Jesús Sol Hernández M (2014) Evaluación in Vivo Del Efecto Cicatrizante de Un Gel a Base de Quitosano Obtenido de Exoesqueleto de Camarón Blanco Litopenaeus Vannamei. Rev Colomb Biotecnol 16(1):45–50
14. Ding H (2012) Modified-release drug products and drug devices. In: Applied biopharmaceutics & pharmacokinetics, pp 567–613
15. Nayak AK, Pal D (2016) Plant-derived polymers: ionically gelled sustained drug release systems. Encycl Biomed Polym 2016:6002–6017. https://doi.org/10.1081/e-ebpp-120050562
16. VR K, S S (2018) Multiple unit particle systems (MUPS), a versatile strategy for controlled drug delivery: focus on formulation and process concerns. J Formul Sci Bioavailab 2018 1(1):1–3. https://doi.org/10.4172/2577-0543.1000106
17. Vishal Y, Satheshkumar S (2019) Current scenario of multiple unit particulate system: a review. Indian J Nov Drug Deliv 11(1):1–12
18. Zhu X, Qi X, Wu Z, Zhang Z, Xing J, Li X (2014) Preparation of multiple-unit floating-bioadhesive cooperative minitablets for improving the oral bioavailability of famotidine in rats. Drug Deliv 21(6):459–466. https://doi.org/10.3109/10717544.2013.879626
19. Alvarez-Lorenzo C, Blanco-Fernandez B, Puga AM, Concheiro A (2013) Crosslinked ionic polysaccharides for stimuli-sensitive drug delivery. Adv Drug Deliv Rev 65(9):1148–1171. https://doi.org/10.1016/j.addr.2013.04.016
20. Liu J, Willför S, Xu C (2015) A review of bioactive plant polysaccharides: biological activities, functionalization, and biomedical applications. Bioact Carbohydrates Diet Fibre 5(1):31–61. https://doi.org/10.1016/j.bcdf.2014.12.001
21. Barclay TG, Day CM, Petrovsky N, Garg S (2019) Review of polysaccharide particle-based functional drug delivery. Carbohydr Polym 221:94–112. https://doi.org/10.1016/j.carbpol.2019.05.067
22. Debele TA, Mekuria SL, Tsai HC (2016) Polysaccharide Based Nanogels in the Drug Delivery System: Application as the Carrier of Pharmaceutical Agents. Mater Sci Eng, C 68:964–981. https://doi.org/10.1016/j.msec.2016.05.121
23. Giri TK (2016) Nanoarchitectured polysaccharide-based drug carrier for ocular therapeutics. Elsevier. https://doi.org/10.1016/B978-0-323-47347-7.00005-7
24. Xie JH, Jin ML, Morris GA, Zha XQ, Chen HQ, Yi Y, Li JE, Wang ZJ, Gao J, Nie SP et al (2016) Advances on bioactive polysaccharides from medicinal plants. Crit Rev Food Sci Nutr 56:S60–S84. https://doi.org/10.1080/10408398.2015.1069255
25. Exner G, Marudova M, Sotirov S, Marinova A, Viraneva A, Pilicheva B, Bodurov I, Vlaeva I, Uzunova Y, Yovcheva T (2019) Multilayered polyelectrolyte structures with potential for intracavity drug delivery systems. Appl Surf Sci 493:620–627. https://doi.org/10.1016/j.apsusc.2019.07.039
26. Stuart MC, de Vries R, Lyklema H (2005) 2 Polyelectrolytes. *Fundam. Interface Colloid Sci* 5(C):1–84. https://doi.org/10.1016/S1874-5679(05)80006-6
27. Shchipunov YA, Postnova IV (2009) Water-soluble polyelectrolyte complexes of oppositely charged polysaccharides. Compos Interfaces 16(4):251–279. https://doi.org/10.1163/156855409X447093
28. Budd PM Polyelectrolytes, In: Allen G, Booth C, Bevington JC, et al. (Eds.), Polymer characterization, comprehensive polymer science: The synthesis, characterization, reactions & applications of polymers, michigan, Vol .1, 968, Elsevier Science & Technology Books

29. Coimbra P, Ferreira P, Alves P, Gil MH (2013) Polysaccharide-based polyelectrolyte complexes and polyelectrolyte multilayers for biomedical applications. Carbohydrates Appl Med 661(2):1–29
30. Sundaramurthy A (2018) Responsive polyelectrolyte multilayer nanofilms for drug delivery applications. Elsevier Ltd. https://doi.org/10.1016/b978-0-08-101997-9.00013-8
31. Racoviţă Ş, Vasiliu S, Popa M, Luca C (2009) Polysaccharides based on micro- and nanoparticles obtained by ionic gelation and their applications as drug delivery systems. Rev Roum Chim 54(9):709–718
32. Patil P, Chavanke D, Wagh M (2012) A review on ionotropic gelation method: novel approach for controlled gastroretentive gelispheres. Int J Pharm Sci 4(Suppl. 4):27–32
33. Lapitsky Y (2014) Ionically crosslinked polyelectrolyte nanocarriers: recent advances and open problems. Curr Opin Colloid Interface Sci 19(2):122–130. https://doi.org/10.1016/j.cocis.2014.03.014
34. Hu Q, Luo Y (2018) Recent advances of polysaccharide-based nanoparticles for oral insulin delivery. Int J Biol Macromol 120:775–782. https://doi.org/10.1016/j.ijbiomac.2018.08.152
35. Jain A, Thakur K, Sharma G, Kush P, Jain UK (2016) Fabrication, characterization and cytotoxicity studies of ionically cross-linked docetaxel loaded chitosan nanoparticles. Carbohydr Polym 137:65–74. https://doi.org/10.1016/j.carbpol.2015.10.012
36. Jayasuriya AC (2017) Production of micro- and nanoscale chitosan particles for biomedical applications. Elsevier. https://doi.org/10.1016/B978-0-08-100230-8.00008-X
37. Crouzier T, Boudou T, Picart C (2010) Polysaccharide-based polyelectrolyte multilayers. Curr Opin Colloid Interface Sci 15(6):417–426. https://doi.org/10.1016/j.cocis.2010.05.007
38. Bantchev G, Lu Z, Lvov Y (2009) Layer-by-layer nanoshell assembly on colloids through simplified wash less process. J Nanosci Nanotechnol 9(1):396–403. https://doi.org/10.1166/jnn.2009.J055
39. Elgadir MA, Uddin MS, Ferdous S, Adam A, Chowdhury AJK, Sarker MZI (2014) Impact of chitosan composites and chitosan nanoparticle composites on various drug delivery systems : a review. J Food Drug Anal, 1–11. https://doi.org/10.1016/j.jfda.2014.10.008
40. Ali A, Ahmed S (2018) International journal of biological macromolecules a review on chitosan and its nanocomposites in drug delivery. Int J Biol Macromol 109:273–286. https://doi.org/10.1016/j.ijbiomac.2017.12.078
41. Kleine-Brueggeney H, Zorzi GK, Fecker T, Gueddari NE, El-Moerschbacher BM, Goycoolea FM (2015) A rational approach towards the design of chitosan-based nanoparticles obtained by ionotropic gelation. Colloids Surf B Biointerfaces 135:99–108
42. Zou X, Zhao X, Ye L, Wang Q, Li H (2014) Preparation and drug release behavior of PH-responsive Bovine Serum albumin-loaded chitosan microspheres. J Ind Eng Chem 21:1389–1397. https://doi.org/10.1016/j.jiec.2014.06.012
43. Ma L, Liu C (2010) Preparation of chitosan microspheres by ionotropic gelation under a high voltage electrostatic field for protein delivery. Colloids Surf B Biointerfaces 75:448–453. https://doi.org/10.1016/j.colsurfb.2009.09.018
44. Nguyen TV, Thu T, Nguyen H, Wang S, Phuong T, Vo K, Nguyen AD (2017) Preparation of chitosan nanoparticles by TPP ionic gelation combined with spray drying, and the and a chitosan nanoparticle—Amoxicillin complex. Res Chem Intermed 43(6):3527–3537. https://doi.org/10.1007/s11164-016-2428-8
45. Antoniou J, Liu F, Majeed H, Qi J, Yokoyama W, Zhong F (2015) Colloids and surfaces A: physicochemical and engineering aspects physicochemical and morphological properties of size-controlled chitosan—Tripolyphosphate nanoparticles. Colloids Surf A Physicochem. Eng Asp 465:137–146
46. Sreekumar S, Goycoolea FM, Moerschbacher, BM, Rivera-rodriguez, GR (2018) Parameters influencing the size of chitosan-TPP nano- and microparticles. Sci Rep, 1–11. https://doi.org/10.1038/s41598-018-23064-4
47. Auwal SM, Zarei M (2017) Improved in vivo efficacy of anti-hypertensive biopeptides encapsulated in chitosan nanoparticles fabricated by ionotropic gelation on spontaneously hypertensive rats. Nanomaterials 7(421), 1–11. https://doi.org/10.3390/nano7120421

Chapter 9
Ionically Gelled Polysaccharide-Based Floating Drug Delivery Systems

Siti Nor Syairah Anis, Ida Idayu Muhamad, Suguna Selvakumaran, Aishah Mohd Marsin, Wen Ching Liew, and Muhamad Elias Alamin Kamaludin

Abstract Polysaccharides from the natural sources are gaining increasing keen from researchers as components of stimuli responsive in drug delivery systems. Polysaccharides are intrinsically biocompatible because the similitude of their structure with many body components and readily degradable in environment by common microorganisms. Ionic polysaccharides can be readily cross-linked to provide hydrogel networks sensitive to a diverse of internal and external variables and thus suitable for switching drug release on-off through various mechanisms for floating drug delivery system inside the gastrointestinal tract. Moreover, the application of polysaccharides gelled into floating drug delivery systems improves the drug bioavailability and patient compliance by increasing the gastric residence time and controlling the drug release. This chapter described the adaptation of drug delivery system by using ionically gelled polysaccharides-based floating drug delivery system inside the gastrointestinal tract.

1 Introduction

The traditional use of excipients in ingredients of drug was to performance as inert vehicles to provided necessary weight, uniformity and volume for the accurate administration of the active ingredient, whereas in modern pharmaceutical dosage forms, they often satisfy multi-functional characters such as modifying release, improvement of the constancy and bioavailability of the active element, improvement of patient adequacy and certify effortlessness ofmanufacture. New and improved excipients remain to be developed to meet the necessities of advanced drug delivery systems

S. N. S. Anis · I. I. Muhamad
IJN-UTM Cardiovascular Engineering Centre (IJN-UTM Cardio Centre), V01, Institute of Human Centered Engineering (iHumEn), Universiti Teknologi Malaysia, Johor Bahru, Johor, Malaysia

I. I. Muhamad (✉) · S. Selvakumaran · A. M. Marsin · W. C. Liew · M. E. A. Kamaludin
Bioprocess & Polymer Engineering Department, School of Chemical and Energy Engineering, Faculty of Engineering, Universiti Teknologi Malaysia, 81310 Johor Bahru, Johor, Malaysia
e-mail: idaidayu@utm.my

A. K. Nayak et al. (eds.), *Ionically Gelled Biopolysaccharide Based Systems in Drug Delivery*, Gels Horizons: From Science to Smart Materials,
https://doi.org/10.1007/978-981-16-2271-7_9

[1]. Basically, drugs delivery systems are described as a drug carrier that reacts with various stimuli to keep the internal environment favourable for the product. The most dynamic drug delivery system is essential in healing numerous illnesses by providing a stable, cost-effective delivery system and effective therapeutic influence to human or animal [2, 3].

There are several routes of drug administration include oral administration, intravenous injection, subcutaneous injection, snorting, inhalation and transdermal administration. Despite tremendous development in drug delivery, oral route of administration has received the most attention and success because the gastrointestinal physiology offers more flexibility in dosage form design than other routes. Oral administration includes swallowing pills, drinking a liquid or eating a substance through the mouth. In this method of administration, most of the absorption of the substance takes place in the gastrointestinal tract [4]. In addition, oral controlled release drug delivery systems have considerable therapeutic benefits such as easiness of administration, patient obedience and flexibility in formulation [5].

The gastrointestinal tract (Fig. 1) is an organ system within humans and other animals which take in food, digests it to extract and absorb energy and nutrients and expels the remaining waste as faeces.

The mouth, oesophagus, stomach and intestines are part of the gastrointestinal tract. Gastrointestinal is an adjective meaning of relating or pertaining to the stomach and intestines. A tract is a collection of related anatomic structures or a series of connected body organs.

Research field in biomedical, nutraceutical and pharmaceutical demonstrated that gastrointestinal tract is the part that is considerate main roles in drug delivery system [6]. The conversions of an oral drug or functional food through the gastrointestinal tract as well as drug release profile exert an excellent encouragement on the effectiveness of medicine and functional food. Drug or functional bioadhesive materials will absorb into stomach wall by active and facilitate transport mechanism [6]. Nevertheless, absorption of drugs through the gastrointestinal tract poses several limitations. For that reasons, researchers develop many strategies for more advance drug delivery system to enhance the absorption of drug into a particular tissues or organs [7]. Other limitations include incapability to restrain and localize the drug delivery system within desired regions of the gastrointestinal tract and the highly variable nature of gastric emptying process [8]. Following on from recent advances, in a time of increased consideration of ionically gelled polysaccharide-based floating drug delivery systems, it is as important as ever to continue the progress by studying different aspects of these systems.

2 Drug Delivery Systems

In drug delivery systems, a problem frequently encountered is an uncontrolled drug delivery, leading to fluctuations in plasma drug level due to its incapability to sustain the dosage form or drug carrier in desired area of the gastrointestinal tract [9]. On the

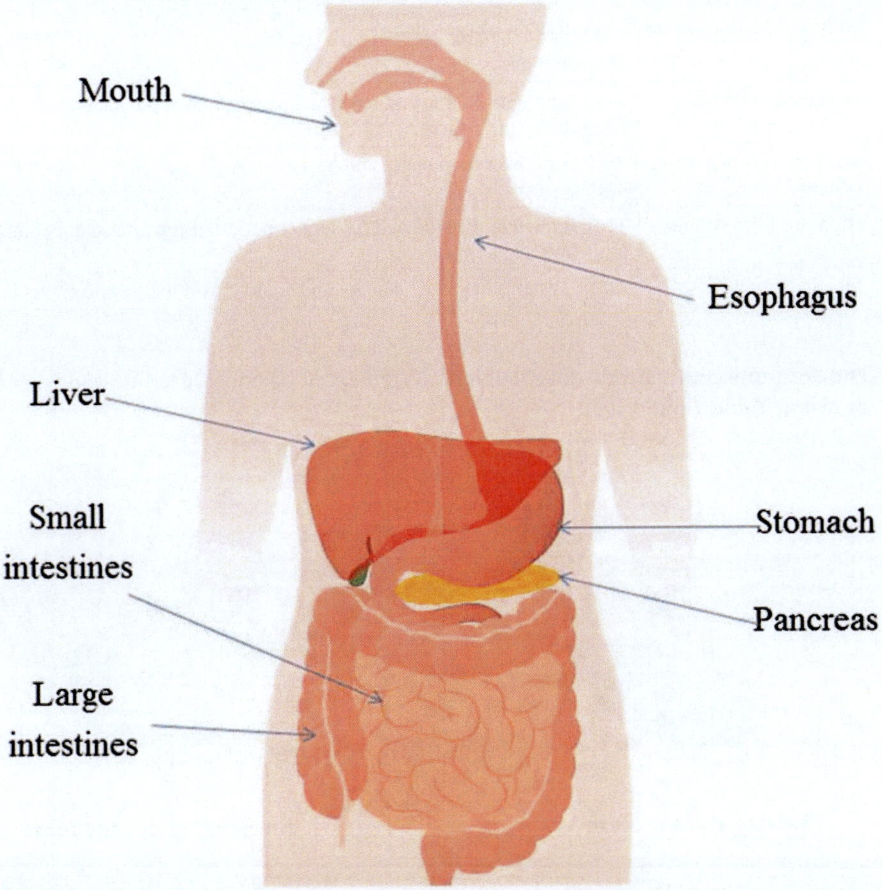

Fig. 1 Gastrointestinal tract

other hand, some dosage form or drug carrier is unstable at various pH conditions, or it may degrade by hostile surroundings in the human body [9–12]. These difficulties may lead to an inadequate drug release from the dosage system and lower the efficacy of administered dosage form and may cause unpredictable bioavailability. Thus, it is necessary to determine drug delivery system that able to extend the residence time of dosage form or carrier in the stomach or somewhere in the upper small intestine until the drug is completely released and thus increase the drug bioavailability. Technological attempts have been prepared in the research and development of rate-controlled oral drug delivery systems to overwhelme physiological difficulties, such as short gastric residence times and unpredictable gastric emptying times [13].

There are several methodologies to elongate the duration of dosage form or drug transporter in the upper part of the gastrointestinal tract including bioadhesive system [14], swelling and expanding systems [15], high-density systems and floating system.

Table 1 Various approaches of drug delivery systems

Systems	Descriptions
Bioadhesive system	They bind with stomach mucosa and hence, enable the localized retention of the system
Swelling and expanding systems	Such systems absorb water and hence, enlarged size
High-density systems	They remain in the stomach for longer period of time, by sedimenting to the folds of stomach
Floating system	These systems have low density and so float over the gastric contents

The description of various drug delivery systems that has been used nowadays was shown in Table 1 and Fig. 2.

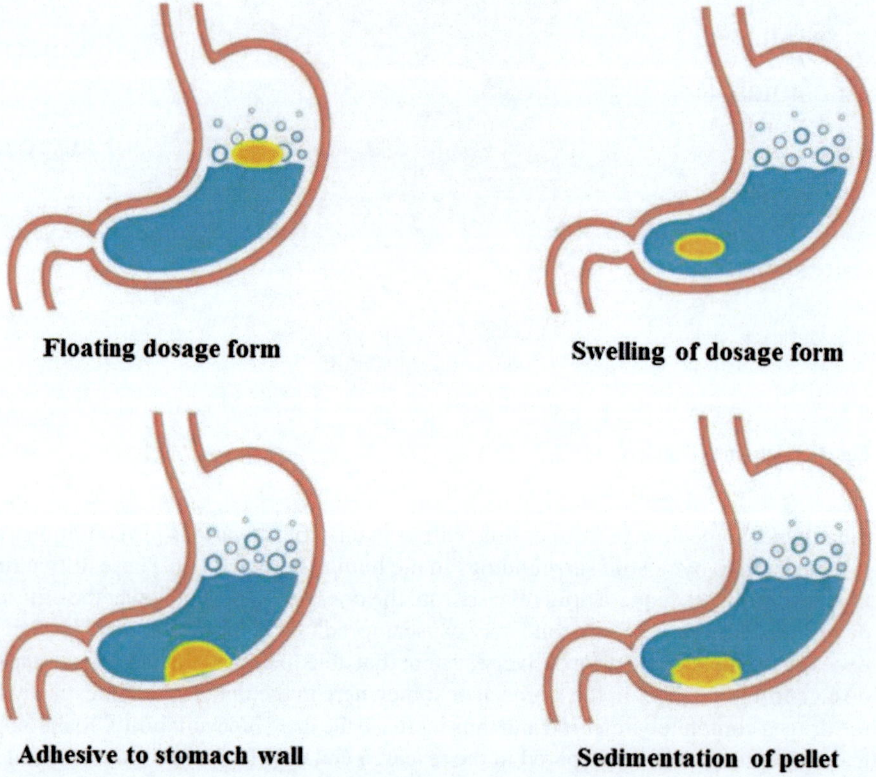

Fig. 2 Different types of drug delivery systems

2.1 Floating Drug Delivery Systems

Among other different approaches of drug delivery system, floating system is the adequate, cost-effective and convenient drug delivery system [17]. Floating system or hydro dynamically balanced system has bulk dense lower than gastric fluids thus remains floating in the stomach without disturbing the gastric emptying rate for an extended period of time in a predetermined manner with the potential for continuous and control release of the drug. After the drug release, the residual system is eliminated from the stomach. This results in an increase in dosage residency time and a better control of the fluctuation in plasma drug concentration [13, 18]. Prolonged residence time along with controlled drug release within the gastrointestinal tract helps to increase drug bioavailability, decrease dosing frequency, improve patient compliance and reduce gastrointestinal side effects [16].

Moreover, multiple-unit floating dosage forms show some advantages over single-unit ones, in terms of uniform dissemination along the gastrointestinal tract, the absence of impairing performance due to failure of a few units, more predictable drug release kinetics and dropping dose-dumping chances [5]. Floating system provides several advantages compared to other conventional drug delivery system [19, 20]. The advantages are as follows:

1. Useful for a drug that is absorbed through the stomach and upper part of the gastrointestinal tract.
2. Helpful for acidic dosage forms that may cause irritation on the stomach wall.
3. Floating system extends drug release over a prolonged period will result in dissipation of drugs in the gastric liquid which would be existing for absorption in the small intestine after emptying of the stomach content. The expectation is the drug that will be fully absorbed from floating dosage forms if it remains in the solution form even at the alkaline pH condition of the intestine.
4. Floating system is useful for unpredictable situations in the body such as stronger intestinal movement and short transit time during diarrhoea that can cause poor absorption.
5. Floating system enhances patient compliance by decreasing dosing frequency.
6. Floating drug delivery enhances bioavailability. Bioavailability is known as the degree of activity or the amount of an administered drug or other substance that becomes available for activity in the target. This leads to a desirable plasma drug concentration that is maintained by the continuous drug release.
7. Better therapeutic effect of short half-life drugs can be achieved.
8. Enhance absorption of drugs which solubilizes in the stomach only.
9. Multiple units of floating system, for example, microspheres release drug equally and no risk of dose dumping.

Despite the fact that floating drug delivery system is necessary in gastrointestinal tract, there are few provocation emerge in evolving floating carrier with pores on the carrier networks. The challenges are to reach optimum drug entrapment similar to native hydrogels and to be able to achieve drug release in a controlled manner.

Furthermore, the carrier should also achieve good gel strength as native hydrogels with optimum porosity and should be safe with excellent bioavailability. Additionally, developed carrier normally has poor physical and mechanical properties that can affect the drug entrapment efficiency and drug release. Thus, modification is needed to enhance these properties.

3 Hydrogels

In recent years, hydrogels have been established as a promising drug carrier in drug delivery application [10–12, 21, 22]. Hydrogels are hydrophilic, a cross-linked polymeric network with three dimensional, and have the ability to imbibe massive amount of water or biological fluid. The water retention properties of hydrogels are due to the existence of hydrophilic groups such as hydroxyl, carboxyl, amino and other chemicals in the polymer forming hydrogel structures [23]. In spite of their high water absorbing ability, hydrogels show good swelling properties which capable to preserve more than 20% of water within its structure without being dissolved when it is in contact with the aqueous surrounding environment [23].

The nature characteristics of hydrogel such as tissue-like and soft physical properties, higher permeability to undersized molecules, release of entrapped molecules in a controlled manner and lower interfacial tension, biocompatible, non-toxic and biodegradable make them an excellent carrier for protein, peptides, drugs and other biological compound [24–27]. In general, hydrogels can change the drug release by changing the gel structure in response to physiological and environmental stimuli including pH, ionic, temperature and electromagnetic radiation. Hydrogels also have the potentiality to preserve the drug from antagonistic environments such as the presence of enzyme and various pH conditions [28, 29]. This notable physiochemical and biological features of hydrogel have led them a promising transporter for drug delivery system. In addition, hydrogels also have widely explored in various biomedical fields, including tissue engineering, biosensor, wound dressing, biomolecule separation and the contact lenses [30, 31].

3.1 *Floating Hydrogels*

Tablets, capsules, granules, laminated films, hollow microspheres beads, powders and hydrogels are samples of advanced buoyant systems [32]. Among these systems, floating hydrogels emerge as a popular and novel approach as drug vehicles in drug delivery system. Interestingly, although floating hydrogel exhibited its floating properties, at the same time, it will demonstrate the characteristics and features of hydrogel such as hydrophilicity, flexibility, versatility, high water absorptivity, biocompatibility and consequently controlled drug release rate. Thus, these properties allow floating hydrogels to be widely used as drug delivery candidates.

Floating hydrogels have become an important field of research because it has ability to increase the residency time of carrier in the stomach and proximal portion of the small intestine, resulting in a sustained and controlled drug released; thereby improving drug bioavailability. The swelling properties of hydrogels prevent their removal via pylorus due to the strong peristaltic waves in stomach. Floating hydrogels make as large as possible the amount of drugs reaching their absorption site in solution and hence ready for absorption [16]. In addition, floating hydrogel results in dissolution of drugs in the gastric fluid. This would then make them accessible for absorption in the small intestine after emptying the stomach content. It is expected that the drugs will be fully absorbed from floating hydrogels if it remains in a solution form even at alkaline pH of the intestinal system [19, 20]. Moreover, the formation of pores within floating hydrogels will facilitate drug release in controlled and sustained manner compared to non-floating hydrogels which hindered the transport of the entrap drug to the surrounding liquid due to compact network structure [10–12].

Polysaccharides are gaining increasing attention as constituents of stimuli-responsive drug delivery systems, predominantly since they can be found in a well characterized and reproducible way from the natural sources. Ionic polysaccharides can be readily cross-linked to render hydrogel networks sensitive to a multiplicity of internal and external variables and thus suitable for switching drug release on-off through diverse mechanisms. Hybrids, composites and grafted polymers can reinforce the responsiveness and broaden the range of stimuli to which polysaccharide-based systems can react [33].

4 Polysaccharide-Based Hydrogels as Biomaterials

In correlating the development of hydrogel from synthetic materials for floating drug purposes, the usage of natural materials is still relevant as the main component to prepare hydrogel. By using advance technology or with the additions of active ingredients, polysaccharide is commonly used in developing hydrogels as biomaterials especially for the usage in medicine area. Polysaccharides are non-toxic, benign to mammalian tissues, which comes from natural resources, and it is a biodegradable polymer. Polysaccharide known to be a large molecule comprises of smaller monosaccharides linked together by glycosidic linkages which form into different homo-(same) or hetero-(different) polysaccharide depends on the connections of monosaccharide as in Fig. 3.

Difference in unity, degree of branching and type and length of the chain differentiates the structure of polysaccharides either forming into a linear or branched polymer. Polysaccharides are not only used as structural material and energy storage system but it was also regarded as important class of polymers to man for their nutritional value as well as industrial utility [34]. Since polysaccharides are obtained from renewable plant in the form of cellulose and starch and from animal resources in the form of glycogen, it could be a niche in becoming the second choices of materials after petroleum-based polymer feedstocks.

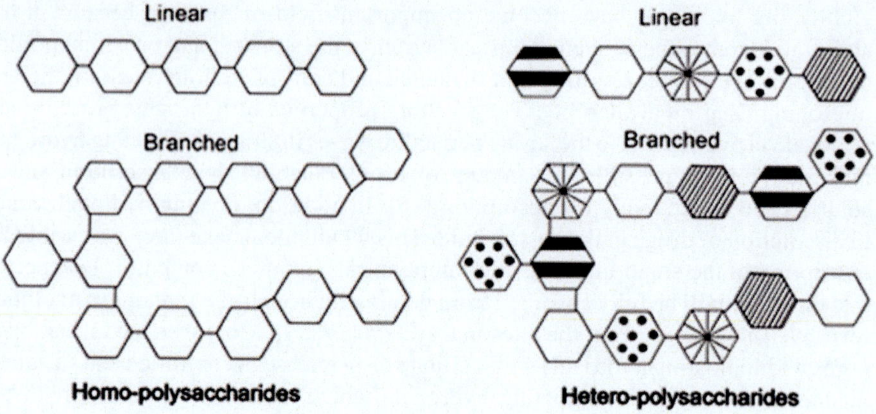

Fig. 3 Homo-polysaccharides and hetero-polysaccharides structure

Polysaccharides become the attention in variety of fields which include pharmaceutics, foods, medicine, tissue engineering, biomedical and cosmetics. Based on their structures, polysaccharides able to have wide variety of functions in nature such as storing energy, sending cellular messages and also providing support to cells and tissues. Polysaccharides in the form of starch and glycogen are functioned as energy storage in plants and mammals, respectively, which easily accessible to the monosaccharides while maintaining a compact structure. In other hand, in order to support cells and tissues, cellulose and chitin usually assembled as a long chain of monosaccharides that form into a fibre and the grouping producing hydrogen bond that strengthen the structure of the material. One of the important source of polysaccharides is algae in the form of phycocolloids such as alginates and carrageenan [35]. Starches and form of hydrocolloids such as plants, pectin, cellulose gam, locust bean gum, from animal such as gelatin and from microbial sources such as xanthan gum and gellan gum are also known as polysaccharides.

The nature of the solvent system characterizes the variety range of polysaccharides based on pH, ionic strength, type of cations and temperatures [36]. Polysaccharide could be divided to neutral, anionic and cationic according to the presence of acid or basic functions [37]. As such ionic polysaccharides are more sensitive to electrolytes (salts) and pH than the neutral polysaccharides. The presence of hydroxyl groups and carboxylic acid groups enables cross-linking of the chains in hydrogel forming ionic gelation as such shown in Fig. 4 for sodium alginate and calcium cross-linking.

Among the present cationic polysaccharide in various industries is chitosan which is used to stabilize the secondary emulsion while examples of commonly used anionic polysaccharides are alginate, xanthan gum, pectins and carrageenan. The attention goes to anionic characters which consist of acidic functions that develops positive charges or brought by cross-linking agent.

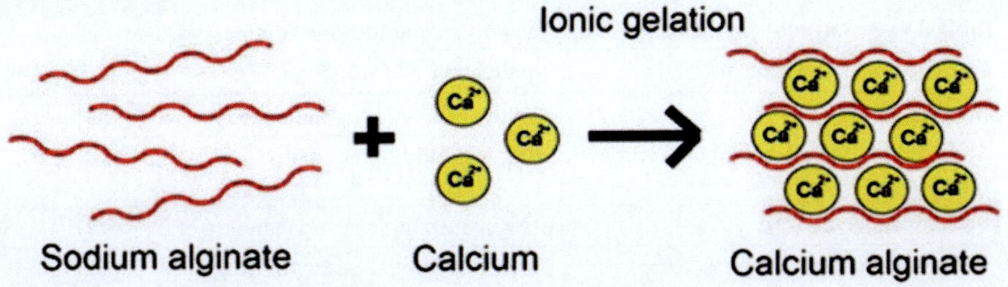

Fig. 4 Ionic gelation cross-linking of hydrogels between anionic polysaccharides (alginate) and divalent cation (Ca^{2+})

4.1 Anionic Polysaccharide

The existence of anionic polysaccharide in biomedical application is capable of self-assembly to form acidic or ionic cross-linked hydrogels with or without chemical functionalization [38]. Anionic polysaccharides will simply form into a gel due to associations between the acid residues and the presence of divalent cations that are most likely to be the favoured characteristic desired in hydrogel application which is known as ionotropic gelation. Anionic polysaccharides are semisynthetic or naturally occurring macromolecules comprising a basic carbohydrate structure that are commonly used in medicine purposes. In floating hydrogel purposes, several anionic polysaccharides were used in gastrointestinal system such as alginate, carrageenan and carboxymethylcellulose. Several researches have studied intense research on floating hydrogel using anionic polysaccharides as in Table 2.

4.1.1 Alginate

The usage of alginate as a gel-forming polymer in developing biomaterials has widely applied especially in pharmaceutical and medicine purposes. Alginate is one of the anionic polysaccharide that can incorporate with divalent cations at room temperature to permitting immobilization under mild and safe conditions [51]. As the resources of alginates come from brown seaweed, the existence of alginate was abundant and becomes one of the renewable resources, biodegradable, biocompatible and non-toxicity. The bioavailability and low cost to extract alginate becomes the main reason of wide usage of alginate in gel development. The complex structure consisting linear binary copolymers of $1 \rightarrow 4$ linked α-*D*-mannuronic acid (M) and β-*L*-glucuronic acid (G) interspersed with MG sequences randomly given the spaces for ion interactions towards cation polysaccharides. G blocks can exchange their Na^+ ions and interact with multivalent cations (commonly Ca^{2+}), and then, the β-*L*-glucuronic acid groups can connect with each other, resulting in a three-dimensional hydrogel network which has pH-sensitive property [52].

Table 2 Several study on the usage of anionic polysaccharides for floating hydrogel

Anionic polysaccharide	Application	References
Sodium alginate	Floating TiO_2/CaAlgphotocatalyst	[39]
Sodium alginate, pectin, hydroxypropyl methylcellulose	Hollow multipolymer matrix	[40]
Sodium alginate	pH-sensitive alginate-Ca^{2+} hydrogel	[41]
Sodium alginate	Floating hydrogel beads for drug gastroretention	[42]
sodium alginate	Polyacrylamide graft–co-polymer-based hydrogel	[43]
HPMC	In situ floating hydrogel for intravesical delivery	[44]
HPMC, carbopol	Floating hydrogel of gastric drug delivery	[45]
HPMC, carbopol	Non-effervescent floating hydrogel	[46]
HPMC, chitosan	Hydrophilic swelling controlled release floating tablet	[47]
HPMC, carbopol	In vitro and in vivo evaluation of floating matrix	[48]
k-carrageenan	Sustained-release floating drug delivery	[49]
k-carrageenan	Sustained-release floating drug delivery	[50]

4.1.2 Hydroxypropyl Methylcellulose

Despite the effectiveness of natural polysaccharides, semisynthetic anionic polysaccharide is also widely used in pharmaceutical purposes to achieve better stability under controlled release of drugs and more compatibility on living organism. Hydroxypropylmethycellulose (HPMC) is one of the semisynthetic polysaccharide available that basically a cellulose ether; prepared by alkali-treated cellulose through the reaction with methyl chloride and propylene oxide. HPMC is also known as propylene glycol ether of methylcellulose. HPMC is commonly used in food industry as additive for emulsifier, thickener and suspender agent in which it becomes the alternative of animal gelatin. Apart from food industry, HPMC is well-known in the applications of drug delivery, dyes, paints, cosmetics, adhesives, coatings, agricultures and textiles due to its hydrophilic, biodegradable and biocompatible properties [53].

4.1.3 Carrageenan

Carrageenan is a family of sulphated, linear polysaccharides that occur in the cell wall and intercellular matrix of red seaweeds. Carrageenan was commonly used as thickening, gelling and stabilizing agents in food industry such as in sauces and dairy products due to its wide availability and exists as one of the polysaccharide family obtained from natural resources. Carrageenan also a highly potential candidate for production of hydrogel to use in various industries based on its non-toxicity, biodegradability and biocompatibility polymer. Carrageenan consists of an alternating linear chain of galactose and anhydro-α-D-galactose unit linked by glycosidic bonds that able to solubilize in hot water and low concentration [54]. Three main varieties of carrageenan exist that differ in degree of sulphation, i.e. kappa-carrageenan, iota-carrageenan and lambda-carrageenan. The primary differences which influence the properties of CG type are the number and position of ester sulphate groups as well as the content of 3,6-anhydro-galactose (3,6-AG). Kappa-carrageenan is mainly used due to its one sulphate group per disaccharide. The usage of kappa-carrageenan was not only restricted to food industry only. In drug delivery system, kappa-carrageenan able to form into oral extended release tablets, polymer mixture, excipient in release of basic drugs, nanoparticles and gelling agent [55]. Carrageenan was used in gastroretention drug delivery since it can be attained by utilizing the swelling property of the carrageenan itself that absorbs fluid from the surrounding environment in a controlled manner, making it float above the gastric contents and remain unaffected by the gastric emptying time [56].

5 Cross-Linker Agent for Floating Hydrogel

In development of hydrogel, cross-linking the polymers with a network structure consists of acidic, basic or neutral monomers will produce a gelling and swelling effect due to the ability to imbibe large amount of water. By cross-linking linear polymers through irradiation or by chemical compound by blowing techniques will develop a superporous and superabsorbent hydrogel with the ability to absorb larger amount of aqueous fluids up to more than few hundred times its own weight [56]. Versatile fluid retention-release character permits superabsorbent hydrogels as smart devices in many fields such as agriculture, hygienic product, wastewater treatment and drug delivery. The desired properties also play important roles in controlled release formulations based on hydrogels. The commonly used cross-linker agent in floating hydrogel application was shown in Table 3. Increase in cross-linker concentration had increased rate of gelation up to a point where cross-linking is restricted due to consequent rise in swelling [57]. As the gelation increases, the entrapment of gas bubble had also increased in gel expanding with greater density and lower percentage of porosity.

Hence, drug release rate decreases which suits the needed potential for drug release in gastrointestinal delivery for a controlled release condition.

Cross-linker agent is the chemical or natural products that form bonds or bridges between two or more polymers to solidify hydrogel and adding more functionality. Since cross-linking is the process to join two or more molecules by a covalent bond, cross-linker agent is required since it consists of two or more reactive ends which are capable or chemically attaching to specific functional groups on the medium. In

Table 3 Study on the incorporation of cross-linker in drug delivery application

Cross-linker	Application	References
Methylenebisacrylamide (BIS)	Poly (acrylamide-co-acrylic acid) superporous hydrogels	[56]
Methylenebisacrylamide (BIS)	Guar gum-g-poly(acrylate) porous superabsorbent hydrogels under different environment	[57]
Methylenebisacrylamide (BIS)	Poly (acrylamide-co-acrylic acid) superporous hydrogels for Vitamin B	[58]
Methylenebisacrylamide (BIS)	Poly (N-isopropylacrylamide) chitosan hydrogel	[59]
Methylenebisacrylamide (BIS)	poly (N-isopropylacrylamide) hydrogel based on freezing polymerization	[60]
Laponite	Poly (N-isopropylacrylamide) free-floating hydrogels	[61]
Laponite	Gellan gum hydrogel	[62]
Genipin	k-Carrageenan/sodium carboxymethyl cellulose floating hydrogel	[63]
Genipin	*N,O*-carboxymethyl chitosan pH-sensitive hydrogel	[64]
Genipin	Kappa-carrageenan/hydroxyethyl cellulose pH-sensitive hydrogels	[65]
Genipin	Chitosan membrane pH-sensitive hydrogel	[66]
Genipin	Kappa-carrageenan/carboxymethyl cellulose beads	[24]

performing floating hydrogel application, only certain cross-linker could be used in order to retain the buoyancy effect and slow release rate of drug containment. Cross-linker agent could be obtained from natural resources or chemically synthesized to synchronize with the biocompatibility of wide drug delivery application.

5.1 Natural Cross-Linker Agent

In compliances with safety and biocompatibility towards human digestion, natural cross-linker agent was used in helping the cross-linking action which occurs in drugs for controlled release purpose in gastrointestinal environment. Genipin, a reagent of plant origin was commonly used for immobilization of monomers by cross-linking action. Genipin is an aglycone derived was obtained from geniposide through treatment with enzyme β-glucosidase, present in G*ardenia jasminoides* extract and are excellent natural cross-linker for proteins, collagen, gelatin, hydrogel and chitosan cross-linking. The usages of genipin prove to improve the mechanical and barrier properties of several proteins medium as shown in Table 3.

Proanthocyanidins collagen or known as grape seed extract which present from fruits, vegetables, nuts, seeds and flowers was mainly used in dentin application. A high structural diversity based on four monomers molecules (catechin, ent-catechin, epicatechin, ent-epicatechin), different types of interflavonoids bonds, and the various lengths of chains, known as degree of polymerization (DP), are unique characteristics of this agent [67].

5.2 *Chemical Cross-Linker Agent*

Since natural cross-linker agent from renewable resources and might be unstable during cross-linking due to various conditions, chemical cross-linker agent would be the best alternative in using a cheaper, more stable, compatible, safe and wide availability. Disuccinimidylsuberate performs as cross-linker agent in developing intermolecular disulphide bonds between oligomers recombinant of bovine and human [68]. The usage of water-soluble carbodiimide (WSC) also affects the gelation of hyaluronic acid which induces intermolecular formation of acid anhydride between carboxyl groups [69].

Glutaraldehyde, a synthetic cross-linking agent has been widely used as fixative agent and reported to improve mechanical properties of various collagen-based tissues. Usage of glutaraldehyde in drugs application could decrease the microsphere size [70], hence increase the density and reduce the rate of release of drugs. However, further applications of glutaraldehyde as the cross-linker agent in drug applications are still under research in order to limit the toxicological effect.

Methylenebisacrylamide (BIS) is a synthetic cross-linking agent specializes in polymerization of acrylamide and polyacrylamide gel formation in which to control the pore size of the gel. BIS could be obtained through the synthesis of acrylamide and formaldehyde in the presence of catalyst [71]. BIS is used in wide applications including wastewater treatment, gel electrophoresis, papermaking, ore processing, tertiary oil recovery and the manufacture of permanent-press fabrics.

Laponite is synthetic smectite clay that consists of phyllosilicates that are characterized by 2:1 crystal composed of layered units in which two tetrahedral silica sheets sandwich one Mg^{2+} containing octahedral sheet. Laponite is degradable under acidic conditions, resulting in controlled release rate for drug entrapment in gastrointestinal conditions [72].

6 Floating Induction Using Pore-Forming Agents

Pore-forming agents or gas generating agents are an essential element to achieve a good floating carrier. Incorporation of carbonates and bicarbonate salt into carrier allows the carrier to constantly float in the stomach and deliver drugs in a controlled manner. The floating behaviour of carrier occurred due to effervescent reactions.

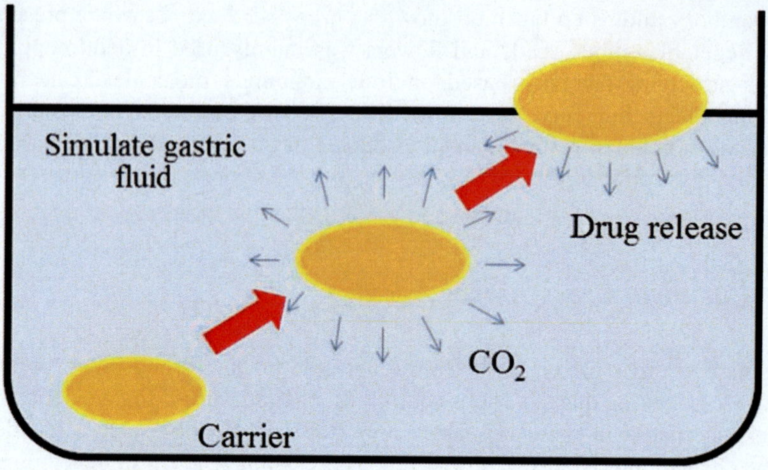

Fig. 5 Schematic diagram of effervescent reaction

Effervescent reactions happened between carbonates or bicarbonates salts and acidic gastric content. Once the carrier is in contact with gastric fluid, carbonates or bicarbonates will effervesce and liberate carbon dioxide (CO_2) which will then get entrapped in the gellified hydrocolloid layer. This reaction makes the density of the carrier reduced and buoyant for a prolonged period with sustained drug release [73]. The schematic diagram of effervescent reaction is shown in Fig. 5.

Sustaining the drug release rate and excellent floatability of dosage form are directly related to the amount and type of pore-forming agents. According to Krishnan et al. [74], there is a direct correlation between the concentration of the pore-forming agents towards the size, weight and pore size as well as the drug release kinetics of the beads. Thus, choosing a suitable pore-forming agents is also very essential for a controlled drug release [74].

6.1 *Sodium Bicarbonate as Pore-Forming Agents*

Sodium bicarbonate ($NaHCO_3$) has frequently been used as a gas-forming agent to develop floating carrier [75–79]. Chen et al. [80] developed floating tablets using sodium bicarbonates ($NaHCO_3$) as gas-forming agents to examine the floating properties and swelling characteristics [80]. It revealed that by adding $NaHCO_3$ makes the tablets reach surfaces within 1 min and sustained buoyancy for more than 24 h. The buoyancy effect shortens the floating lag time further because gas bubbles are formed after reacting with acidic medium. They found that addition of $NaHCO_3$ could increase the floating capabilities (floating lag time and floating duration). There are also some reports on potassium carbonates and calcium carbonates as pore-forming agents in developing floating carriers [74, 81].

6.2 Calcium Carbonate as Pore-Forming Agents

Calcium carbonate is one of the common pore-forming agents that is generally used in numerous applications. Calcium carbonate is a chemical compound with the chemical formula $CaCO_3$. It is formed by carbon, oxygen and calcium. This pore-forming agents extensively been utilized in industrial of application, health and dietary application, environmental application and food industry.

In medical application, $CaCO_3$ is used as a low-priced dietary calcium supplement or gastric antacid. It is also employed as a phosphate binder for the treatment of hyperphosphatemia (chronic renal failure) and as filler for tablets in pharmaceutical industry [82]. In food industry, $CaCO_3$ acts as food additive and is used as acidity regulator, anticaking agents and stabilizer, dietary calcium, soy milk and almond milks [83] and as firming agents in many canned or bottled vegetable products.

In drug delivery system, $CaCO_3$ has wide application due to its high availability, low cost, safety, biocompatibility, pH sensitivity and slow biodegradability. Hence, it can be used as pore-forming agents for developing a floating carrier in drug delivery system. $CaCO_3$ in carrier will react with gastric fluid and release CO_2 and permeates pore on hydrogels matrix causing the hydrogel to become less dense than the fluid, making it to remain afloat for a longer time and release drug in a controlled manner [81, 84].

Moreover, Abou el Ela and co-workers [85] synthesized floating beads using $CaCO_3$as gas-forming agents to make a sustained release of ketorolac tromethamine in the stomach [85]. Yellanki and Neralla (2010) utilize $CaCO_3$as gas-forming agents to develop floating alginates beads for stomach-specific drug delivery. It was found that $CaCO_3$amount gives significant effect on size analysis, morphology test, drug entrapment efficiency, buoyancy study and in vitro release [81]. There are several research papers on $CaCO_3$as pore-forming agents for floating system [84, 86].

Choi et al. (2002) prepared floating alginate beads to investigate the effect of two different CO_2 gas-forming agents ($CaCO_3$ and $NaHCO_3$). Riboflavin was used as a model drug. The results showed that the drug release rate was increased consistently with the addition of $CaCO_3$ and NaHCO3. They concluded that $CaCO_3$ is an excellent CO_2 gas-forming agent compared to $NaHCO_3$ in alginate bead preparations [87].

7 Application of Nanoparticles

Nanoparticles are known as small particle varied in size from 10 to 1000 nm [88]. Nowadays, the application of nanoparticles in floating drug system delivery also gives a great deal for a efficient drug delivery since it offers site-specific and time controlled drug delivery [65, 89]. Moreover, nanoparticles have the potential to revolutionize drug delivery system because of their properties such as enhancing bioavailability, providing targeted drug delivery, prolonging the drug effect in targeted site,

Table 4 Advantages and disadvantages of nanoparticles in drug delivery application

Advantages	Disadvantages
1. Surface properties and particle size of nanoparticles can be easily manipulated in order to achieve both active and passive drug targeting after parenteral administration	1. Nanoparticles in small size and large surface area can lead to particle aggregation Ref [92]
2. Nanoparticles can control and sustain the drug release, altering the drug distribution in organ, and subsequent clearance of the drug so as to achieve an increase in drug therapeutic efficacy and reduction in side effects	
3. Relatively high drug loading and there is no chemical reaction when drug is incorporated into the system	
4. Site-specific drug targeting can be prepared by attaching targeting ligands to surface of particles or use of magnetic guidance	

improving the stability of therapeutic agents against chemicals or enzymatic degradation and also emerged as potent effective agents against infections due to its antibacterial effect [90, 91]. Table 4 showed the advantages and disadvantages of nanoparticles in drug delivery application.

In the application in floating drug delivery, some ionically gelled polysaccharides may endow nanocarriers with external characteristics that capable to control the interactions with the main elements they will enter into contact during absorption and biodistribution, namely mucosa, blood and target cells. Biomimicking the surface of eukaryotic cells, bacteria and viruses, polysaccharides can facilitate the recognition and binding to targeted surfaces, while scaping from opsonisation and accompaniment activation. Consequently, integration of the polysaccharides features in nano/micro/macro-hydrogel systems is particularly attractive to attain novel biocompatible, responsive and even targetable drug delivery system, appropriate to be managed via virtually any route and a promising drug carrier as well as an antibacterial agent [33, 93, 94].

8 Controlled Release of Drug Delivery System

Controlled released drug delivery system is one of the effective novel drug delivery system introduced to overcome the drawback of fluctuating drug level associated with conventional dosage forms. Controlled drug delivery system refers to a way to deliver drug at a predetermined rate, for locally or systematically, for a specific period of time. This system proposes great advantages over conventional dosage forms as follow [95]:

1. Be applied for localized drug delivery.
2. Diminish in vivo fluctuation of drug concentrations thus maintain its desired drug concentration for prolonged duration of treatment.
3. Maximize the efficacy-dose relationship.
4. Reduce the dosage frequency to improve the patient compliance.
5. Decrease the adverse side effects.

While these advantages can be significant, the potential disadvantages cannot be neglected such as dose dumping, chemical, physical and dissolution storage stability problem as well as reduced potential for accurate dose adjustment. The attempts to extend the residence time of drugs in small intestine even more challenging to improve drugs absorption efficiency. Hence, utilization of ionically gelled polysaccharide-based floating drug delivery systems approaches has been done to overcome this issue.

8.1 Mechanisms of Burst Release in Polymeric Hydrogels

The main major challenges in the development of polymeric delivery vehicles are the initial burst release of the drug. Burst release, known as an initial large bolus drug, is released immediately upon placement in the release medium, which is before the release rate reaches a stable profile [96, 97]. Generally, burst release happens due to various reasons including (i) poor distribution of drug within the hydrogels networks during formation, drying and storage, (ii) drug loosely attached on the surface of hydrogels, (iii) the nature and molecular weight of the polymer and drug and (iv) sample geometry and morphology [98–101]. Although burst release is helpful for some applications such as dermal drug delivery for the penetration of a drug [98], but in many cases, it provided negative phenomenon.

Many approaches have been tried to eliminate the initial burst release including changing polymer formulations and morphology [102], surface coating [103, 104] and surface cross-linking [105]. Ying et al. (2008) reported that after cross-linking by glutaraldehyde, the burst release of doxorubicin (DOX) from nanostructured lipid carriers (NLC) could be reduced, in comparison with non-cross-linked NLC [106]. In another study, Hezaveh and Muhamad [13] used non-toxic natural genipin cross-linker on kappa-carrageenan/polyvinyl alcohol to minimize burst release effect in β-carotene release. It was found that cross-linker can minimize burst release of active material than the native films. This is because different genipin amount had changed the morphology of the polymeric network which can eliminate the burst effect. By modifying the drug distribution in polymers network, the initial burst release can be controlled [10–12].

8.2 Mechanisms of Ionic Polysaccharides

Neutral and polyelectrolyte polysaccharides can be endowed with additional ionisable groups by means of replacement of some hydroxyl groups with ionic moieties, by grafting of polymerizable ionic monomers or preformed polymers or can come from the physically and chemically cross-linker used to prepare the networks. Cationic polysaccharides are prone to swell in acid pH and to shrink at neutral/alkaline pH. Thus, networks showing swelling controlled release deliver faster the drug at acid pH.

On the other hand, the opposite effect is observed in the case of anionic polysaccharides. Ions help to regulate the swelling degree through osmotic effects but also increasing the cross-linking density, thus tuning the release rate. Affinity-controlled mechanisms can also occur by means of ionic interactions between the polysaccharides and oppositely charged drug molecules. In that case, the pH induced swelling might not play any character in the regulator of drug release. These features make polysaccharides appropriate for site-specific oral delivery, predominantly for release at the colon. In this sense, resistance to degradation by upper gastrointestinal tract enzymes and susceptibility to large intestine enzymes may synergistically contribute to efficient site-specific release [33].

The swelling mechanisms of ionic gelled polysaccharide inside gastrointestinal tract as predictable relates to the repulsive effect of negative charges of carboxylate groups. The hydrogel swelling to a larger size and floating in gastric fluids can further extend stomach retention. Hydroxyethyl cellulose (HEC) is a non-ionic and water-soluble polymer widely used in pharmaceutical formulations. Simply incorporating negatively charged sodium carboxymethyl cellulose (NaCMC) in the HEC matrix might expand the gel matrix of HEC via repulsive forces to a greater extent to enhance its gastro retentive ability. Addition of sodium chloride (NaCl) to the media affected the solubility of the drugs, and also their gelling behaviours, resulting in different mechanisms for controlling a drug's release. This development possesses proper swelling extents and desired floating periods with sustained-release characteristics [107].

9 Biocompatibility Studies

Ionic polymeric hydrogels are extensively applied as drug transport in drug delivery system. Hence, hydrogels as drug carrier must be biocompatible and non-toxic in order to be safely used. Biocompatibility is also recognized as the ability of materials to execute with an applicable host response in particular application deprived of any harmful and side effect [108].

Generally, polymeric hydrogels are biocompatible, biodegradable and demonstrated an exceptional biological activity. Nevertheless, the properties of natural polymers such as low mechanical strength and high rates of degradation often consequences in their usage in combination with nanoparticles or in subsequent cross-linking reactions to decrease the degradation rates. Additionally, unreacted monomers, initiators and cross-linker can leach out throughout application and cause toxicity to human body [23, 109].

It is obligatory to carry out a number of tests on hydrogels to evaluate their biological tolerance and biocompatibility. The test may comprise *in vivo* analysis using animals such as mice and in vitro analysis using cell line. In vivo test is usually expensive and involves regulatory concerns. Due to this, in vitro testing is regularly performed to characterize newly developed biomaterial. In vitro tests for biocompatibility analysis comprise cytotoxicity tests of the feature of the material in the presence of live host cells and can commonly be done in two ways; direct method and indirect method (MTT assay).

9.1 Kappa-Carrageenan (κ-Carrageenan) Biocompatibility

Exclusive characteristics such as biocompatibility, non-toxicity, biodegradability and environmental sensitivity are the determining factors in materials selection for the floating drug delivery system. Among all biopolymers, κ-carrageenan stands out as great candidate in biomedical applications as it fulfils all these characteristics with good hydrophilicity and gel-forming ability too. Kappa-carrageenan is dietary fibres that does not degrade within the upper GI tract. However, when it is in the colon, it disintegrates and delivers the active ingredients. With this feature, κ-carrageenan is applied as coating agent in oral drug delivery for specific delivery purpose. κ-carrageenan is also favoured for their biological properties like anti-tumor activity, anticoagulant activity, immunomodulatory effect and inflammatory responses [110, 111].

In vitro cytotoxicity evaluation of κ-carrageenan hydrogel using L929 mouse fibroblast cell line showed that κ-carrageenan did not affect L929 metabolic activity significantly. The in vivo evaluation indicated κ-carrageenan induced a low inflammatory response, suggesting κ-carrageenan is biocompatible [112]. Besides, Suguna and co-workers also reported, κ-carrageenan-based floating hydrogel formulated using $CaCO_3$ as pore-forming agents with genipin and magnesium oxide nanoparticles proved its excellent dual functioned carrier for gastrointestinal tract with controlled drug release and reduced initial burst. These follow along with the ability to inhibit the growth of pathogenic bacteria in gastrointestinal tract, biocompatibility, non-toxicity, and it is safe to use [49, 63].

10 Conclusion

As a conclusion, ionically gelled polysaccharide-based floating drug delivery systems approach clearly exhibited that residence time can be prolonged over long period.

Thus, cross-linking ionic mechanisms and incorporation of nanoparticles in floating hydrogels are expected to control the drug release rate with minimal initial burst release and enhanced physical and mechanical properties. Furthermore, the presence of nanoparticles in floating hydrogels is estimated to show excellent antibacterial effect towards pathogenic bacteria in gastrointestinal tract.

Acknowledgements The authors would like to thank the Ministry of Science, Technology and Innovation, Malaysia, Ministry of Higher Education Malaysia and Research Management Centre of UniversitiTeknologi Malaysia (UTM) for the support (grant no. R.J130000.7113.04E94) of this study.

References

1. Krishna LNV, Kulkarni PK, Dixit M, Lavanya D, Raavi PK (2011) Brief introduction of natural gums, mucilages and their applications in novel drug delivery systems-a review. Int J Drug Formul Res 2(6):54–71
2. Bajpai SK, Sonkusley J (2002) Hydrogels for oral drug delivery of peptides: synthesis and characterization. J Appl Polym Sciences 83:1717–1729
3. Graham NB, Mc-Neil ME (1984) Hydrogels for controlled drug delivery. Biomat 5(1):27–36
4. Dubey J, Verma N (2013) Floating drug delivery system: A review. Int J Pharm Sci Res 4(8):2893
5. Nayak AK, Pal D, Malakar J (2013) Development, optimization, and evaluation of emulsion-gelled floating beads using natural polysaccharide-blend for controlled drug release. Polym Eng Sci 53(2):238–250
6. Borase CB (2012) Floating systems for oral controlled release drug delivery. Int J Appl Pharm 4(2):1–13
7. Dhole AR, Gaikwad PD, Bankar VH, Pawar SP (2011) A review on floating multiparticulate drug delivery system-A novel approach to gastric retention. Int J Pharm Sci Rev Res 2:205–211
8. Sharma M (2013) A Review of Floating Drug Delivery Systems. Asian Journal of Biomedical and Pharmaceutical Sciences 3(24):1
9. Singh BM, Kim KH (2000) Floating drug delivery systems: an approach to oral controlled drug delivery *via* gastric retention. J Controlled Release 63:235–259
10. Hezaveh H, Muhamad II (2013) Effect of MgO nanofillers on burst release reduction from hydrogel nanocomposites. J Mater Sci: Mater Med
11. Hezaveh H, Muhamad II (2013) Modification and swelling kinetic study of kappa-carrageenan-based hydrogel for controlled release study. J Taiwan Inst Chem Eng 44:182–191
12. Hezaveh H, Muhamad II (2013) Controlled drug release *via* minimization of burst release in pH-response kappa-carrageenan/polyvinyl alcohol hydrogels. Chem Eng Res Des 9(1):508–519
13. Shah SH, Patel JK, Patel NV (2009) Stomach specific floating drug delivery system: A review. Int J Pharm Tech Res 1(3):623–633
14. Tangri T, Khurana S, Madhav SNV (2011) Mucoadhesive drug delivery: mechanism and methods of evaluation. Int J Pharma Bio Sci 2(1)
15. Omidian H, Park K (2008) Swelling agents and devices in oral drug delivery. J Drug Deliv Sci Technol 18(2):83–93
16. Dolas TR, Hosmani A, Bhandari A, Kumar B, Somvanshi S (2011) Novel sustained release gastroretentive drug delivery system: a review. Int J Pharma Res Dev 2(11):26–41
17. Swamy PV, Bhosale UV, Hiremath SN, Purohit S (2007) Recent trends in drug research: gastroretentive floating drug delivery systems. Asian J Pharm 1(2–3):129–135

18. Dixit N (2011) Floating drug delivery system. J Curr Pharm Res 7(1):6–20
19. Chordiya MA, Senthilkumaran K, Gangurde HH, Tamizharasi S (2011) Floating drug delivery system: a versatile approach for gastric retention. Int J Front Res 1(3):96–112
20. Mayavanshi AV, Gajjar SS (2008) Floating drug delivery systems to increase gastric retention of drugs: a review. Res J Pharm Technol 1(4)
21. Salgueiro AM, Daniel-Da-Silva AL, Fateixa S, Trindade T (2013) K-Carrageenan hydrogel nanocomposites with release behaviour mediated by morphological distinct Au nanofillers. Carbohyd Polym 91:100–109
22. Mohan YM, Vimala K, Thomas V, Varaprasad K, Sreedhar B, Bajpai SK, Raju KM (2010) Controlling of silver nanoparticles structure by hydrogel networks. J Colloid Interface Sci 342:73–82
23. Peppas NA, Bures P, Leobandung W, Ichikawa H (2000) Hydrogels in pharmaceutical formulations. Eur J Pharm Biopharm 50:27–46
24. Muhamad II, Fen LS, Hui NH, Mustapha NA (2011) Genipin-cross-linked kappa-carrageenan/carboxymethyl cellulose beads and effects on beta-carotene release. Carbohyd Polym 83(3):1207–1212
25. Ratner BD, Hoffman AS (1976) Synthetic hydrogels for biomedical applications. In: Andrade JD (Ed.), Hydrogels for medical and related applications. ACS Symposium Series, No. 31, American Chemical Society, Washington, DC, pp 1–36
26. Ratner BD, Williams DF (1981) Biocompatibility of clinical implant materials. CRC, Boca Raton, FL
27. Ratner BD, Hoffman AS, Andrade JD (1976) Hydrogels for medical and related applications. American Chemical Society, Washington, DC, p 1
28. Nho YC, Park SE, Kim HI, Hwang TS (2005) Oral delivery of insulin using pH sensitive hydrogels based on polyvinyl alcohol grafted with acrylic acid/methacrylic acid by radiation. Nucl Instrum Methods Phys Res B 236:283–288
29. Qiu Y, Park K (2001) Environment-Sensitive hydrogels for drug delivery. Adv Drug Deliv Rev 53:321–339
30. Slaughter BV, Khurshid SS, Fisher OZ, Khademhosseini A, Peppas NA (2009) Hydrogels in regenerative medicine. Adv Mat 21:3307–3329
31. Yaszemski M, Trntolo D, Lewandrowski KU, Hasirci V, Altobelli D, Wise D (Eds) (2004) Tissue engineering and novel delivery systems. CRC Press
32. Bhardwaj V, Nirmala HS (2013) Floating drug delivery system: A review. Pharmacophore 4(1):26–38
33. Alvarez-Lorenzo C, Blanco-Fernandez B, Puga AM, Concheiro A (2013) Crosslinked ionic polysaccharides for stimuli-sensitive drug delivery. Adv Drug Deliv Rev 65(9):1148–1171
34. Partain EM (2000) Industrially important polysaccharides. In: Applied polymer science: 21st century. Pergamon, pp 303–323
35. Udayan A, Arumugam M, Pandey A (2017) Nutraceuticals from algae and cyanobacteria. In: Algal green chemistry. Elsevier, pp 65–89
36. BeMiller JN (2019) Polysaccharides: Properties. Carbohydrate chemistry for food scientists, pp 103–157
37. Souguir Z, About-Jaudet E, Picton L, Le Cerf D (2015) Anionic Polysaccharide hydrogels with charges provided by the polysaccharide or the crosslinking agent. Drug Deliv Lett 2(4):240–250
38. Tai C, Bouissil S, Gantumur E, Carranza MS, Yoshii A, Sakai S, Delattre C (2019) Use of anionic polysaccharides in the development of 3D bioprinting technology. Appl Sci (Switzerland) 9(2596):1–13
39. Dalponte I, de Sousa BC, Mathias AL, Jorge RMM (2019) Formulation and optimization of a novel TiO_2/calcium alginate floating photocatalyst. Int J Biol Macromol 137:992–1001
40. Auriemma G, Cerciello A, Sansone F, Pinto A, Morello S, Aquino RP (2018) Polysaccharides based gastroretentive system to sustain piroxicam release: Development and *in vivo* prolonged anti-inflammatory effect. Int J Biol Macromol 120:2303–2312

41. Cong Z, Shi Y, Wang Y, Wang Y, Niu J, Chen N, Xue H (2018) A novel controlled drug delivery system based on alginate hydrogel/chitosan micelle composites. Int J Biol Macromol 107(Part A):855–864
42. Murata Y, Sasaki N, Miyamoto E, Kawashima S (2000) Use of floating alginate gel beads for stomach-specific drug delivery. Eur J Pharm Biopharm 50(2):221–226
43. Tripathi R, Mishra B (2012) Development and evaluation of sodium alginate–polyacrylamide graft–co-polymer-based stomach targeted hydrogels of famotidine. AapsPharmscitech 13(4):1091–1102
44. Lin T, Wu J, Zhao X, Lian H, Yuan A, Tang X, Hu Y (2014) In situ floating hydrogel for intravesical delivery of adriamycin without blocking urinary tract. J Pharm Sci 103(3):927–936
45. Li S, Lin S, Daggy BP, Mirchandani HL, Chien YW (2003) Effect of HPMC and Carbopol on the release and floating properties of Gastric Floating Drug Delivery System using factorial design. Int J Pharm 253(1–2):13–22
46. Acharya S, Patra S, Pani NR (2014) Optimization of HPMC and carbopol concentrations in non-effervescent floating tablet through factorial design. Carbohyd Polym 102(1):360–368
47. Verma A, Sahu AK, Singh SK (2012) Preparation of Hydrophilic Swelling Controlled-Release Floating Matrix Tablets Containing HPMC and Chitosan. International Journal of Pharmacy and Pharmaceutical Sciences 4(1):1–7
48. Xu X, Sun M, Zhi F, Hu Y (2006) Floating matrix dosage form for phenoporlamine hydrochloride based on gas forming agent: *In vitro* and *in vivo* evaluation in healthy volunteers. Int J Pharm 310(1–2):139–145
49. Selvakumaran S, Muhamad II, AbdRazak SI (2016) Evaluation of kappa carrageenan as potential carrier for floating drug delivery system: Effect of pore forming agents. Carbohyd Polym 135:207–214
50. Onyishi IV, Chime SA, Egwu E (2013) Application of κ-carrageenan as a sustained release matrix in floating tablets containing sodium salicylate. African Journal of Pharmacy and Pharmacology 7(39):2667–2673
51. Leong JY, Lam WH, Ho KW, Voo WP, Lee MFX, Lim HP, Chan ES (2016) Advances in fabricating spherical alginate hydrogels with controlled particle designs by ionotropic gelation as encapsulation systems. Particuology 24:44–60
52. Mohamadnia Z, Zohuriaan-Mehr MJ, Kabiri K, Jamshidi A, Mobedi H (2008) Ionically cross-linked carrageenan-alginate hydrogel beads. J Biomater Sci Polym Ed 19(1):47–59
53. Deshmukh K, Basheer Ahamed M, Deshmukh RR, Khadheer Pasha SK, Bhagat PR, Chidambaram K (2016) Biopolymer composites with high dielectric performance: interface engineering. In: Biopolymer composites in electronics, pp 27–128
54. Tecante A, Núñez C (2012) Solution properties of κ-carrageenan and its interaction with other polysaccharides in aqueous media. Rheology, pp 241–264
55. Li L, Ni R, Shao Y, Mao S (2014) Carrageenan and its applications in drug delivery. Carbohyd Polym 103(1):1–11
56. Chavda H, Patel C (2011) Effect of crosslinker concentration on characteristics of superporous hydrogel. Int J Pharm Investig 1(1):17
57. Chandrika KP, Singh A, Rathore A, Kumar A (2016) Novel cross linked guar gum-g-poly(acrylate) porous superabsorbent hydrogels: Characterization and swelling behaviour in different environments. Carbohyd Polym 149:175–185
58. Bajpai SK, Bajpai M, Sharma L (2007) Prolonged gastric delivery of vitamin B2 from a floating drug delivery system: An *in vitro* study. Iran Polym J (Engl Ed) 16(8):521–527
59. Chen KS, Ku YA, Lee CH, Lin HR, Lin FH, Chen TM (2005) Immobilization of chitosan gel with cross-linking reagent on PNIPAAm gel/PP nonwoven composites surface. Mater Sci Eng, C 25(4):472–478
60. Xue W, Hamley IW, Huglin MB (2002) Rapid swelling and deswelling of thermo reversible hydrophobically modified poly(N-isopropylacrylamide) hydrogels prepared by freezing polymerisation. Polymer 43(19):5181–5186

61. Gregory J, Riasi MS, Cannell J, Arora H, Yeghiazarian L, Nistor V (2014) Remote-controlled peristaltic locomotion in free-floating PNIPAM hydrogels. J Appl Polym Sci 131(20):2–7
62. Pacelli S, Paolicelli P, Avitabile M, Varani G, Di Muzio L, Cesa S, Casadei MA (2018) Design of a tunable nanocomposite double network hydrogel based on gellan gum for drug delivery applications. Eur Polymer J 104(March):184–193
63. Selvakumaran S, Muhamad II (2015) Evaluation of kappa carrageenan as potential carrier for floating drug delivery system: Effect of cross linker. Int J Pharm 496(2):323–331
64. Chen SC, Wu YC, Mi FL, Lin YH, Yu LC, Sung HW (2004) A novel pH-sensitive hydrogel composed of N, O-carboxymethyl chitosan and alginate cross-linked by genipin for protein drug delivery. J Controlled Release 96(2):285–300
65. Hezaveh H, Muhamad II (2012) Effect of natural cross-linker on swelling and structural stability of kappa-carrageenan/hydroxyethyl cellulose pH-sensitive hydrogels. Korean J Chem Eng 29(11):1647–1655
66. Mi FL, Shyu SS, Peng CK (2005) Characterization of ring-opening polymerization of genipin and pH-dependent cross-linking reactions between chitosan and genipin. J Polym Sci, Part A: Polym Chem 43(10):1985–2000
67. Castellan CS, Bedran-Russo AK, Antunes A, Pereira PNR (2013) Effect of dentin biomodification using naturally derived collagen cross-linkers: One-year bond strength study. Int J Dent
68. Guttmann RP, Erickson AC, Johnson GVW (2002) Self-association: Stabilization with a chemical cross-linker and modulation by phosphorylation and oxidation state. J Neurochem 64(3):1209–1215
69. Maleki A, Kjøniksen AL, Nyström B (2007) Characterization of the chemical degradation of hyaluronic acid during chemical gelation in the presence of different cross-linker agents. Carbohyd Res 342(18):2776–2792
70. Genta I, Costantini M, Asti A, Conti B, Montanari L (1998) Influence of glutaraldehyde on drug release and mucoadhesive properties of chitosan microspheres. Carbohyd Polym 36(2–3):81–88
71. Avşar A, Gökbulut Y, Ay B, Serin S (2017) A novel catalyst system for the synthesis of N, N′-Methylenebisacrylamide from acrylamide. Des Monomers Polym 20(1):434–440
72. Tomás H, Alves CS, Rodrigues J (2018) Laponite®: A key nanoplatform for biomedical applications? Nanomedicine: Nanotechnol Biol, Med 14(7):2407–2420
73. Ahmed A, Goyal NR, Sharma PK (2014) Effervescent floating drug delivery system. Glob J Pharmacol 8(4):478–485
74. Krishnan V, Sasikumar S, Dass FP, Vijayaraghavan R (2010) Effect of pore forming agents on the physical characteristics and release kinetics of levofloxacin hemihydrate from floating alginate drug delivery system—an *in vitro* study. Trends BiomaterArtif Organs 24(3):139–145
75. Park YJ, Liang J, Yang Z, Yang VC (2001) Controlled release of clot-dissolving tissue-type plasminogen activator from a poly(L-Glutamic Acid) semi-interpenetrating polymer network hydrogel. J Controlled Release 75:37–44
76. Baumgarter S, Kristl J, Vrecer F, Vodopivec P, Zorko B (2000) Optimisation of floating matrix tablets and evaluation of their gastric residence time. Int J Pharm 195:125–135
77. Chen J, Park K (2000) Synthesis of fast-swelling, superporous sucrose hydrogels. Carbohyd Polym 41:259–268
78. Mandel KG, Daggy BP, Brodie DA, Jacoby HI (2000) Alginate-raft formulations in the treatment of heartburn and acid reflux. Aliment Pharmacol Ther 14:669–690
79. Zaniboni HC, Fell JT, Collett JH (1995) Production and characterisation of enteric beads. Int J Pharm 125:151–155
80. Chen Y-C, Hoa HO, Lee TY, Sheu MS (2013) Physical characterizations and sustained release profiling of Gastroretentive drug delivery systems with improved floating and swelling capabilities. Int J Pharm 441:162–169
81. Yellanki SK, Nerella NK (2010) Stomach specific drug delivery of riboflavin using floating alginate beads. Int J Pharm Pharm Sci 2(2):160–163

82. Lieberman HA, Lachman L, Schwartz JB (1990) Pharmaceutical dosage forms: tablets. Dekker, New York, p 153
83. Zhao Y, Martin BR, Weaver CM (2005) Calcium Bioavailability of Calcium Carbonate Fortified Soymilk is Equivalent to Cow's Milk in Young Women. J Nutr 135(10):2379–2382
84. Sriamornsak P, Sungthongjeen S, Puttipipatkhachorn S (2007) Use of pectin as a carrier for intragastric floating drug delivery: Carbonate salt contained beads. Carbohyd Polym 67(3):436–445
85. Abou El Ela AESF, Hassan MA, El-Maraghy D (2014) Ketorolac tromethamine floating beads for oral application: characterization and *In Vitro/In Vivo* evaluation. Saudi Pharmaceutical J 22:349–359
86. Pahwa R, Bhagwan S, Kumar V, Kohli K (2010) Available online through role of natural polymers in the development of floating drug delivery systems. J Pharm Res 3(6):1312–1318
87. Choi BY, Park HJ, Hwang SJ, Park JB (2002) Preparation of alginate beads for floating drug delivery system: effects of CO_2 gas-forming agents. Int J Pharm 239(1–2):81–91
88. Jong WHD, Borm PJA (2008) Drug delivery and nanoparticles: applications and hazards. Int J Nanomed 3(2):133–149
89. Hamidi M, Azadi A, Rafiei P (2008) Hydrogel nanoparticles in drug delivery. Adv Drug Deliv Rev 60:1638–1649
90. Salouti M, Ahangari A (2014) Nanoparticle based drug delivery systems for treatment of infectious diseases. Application of Nanotechnology in Drug Delivery
91. Rai RV, Bai JA (2011) Nanoparticles and Their Potential Application as Antimicrobials. Science against microbial pathogens: communicating current research and technological advances, 197–209
92. Mohanraj VJ, Chen Y (2006) Nanoparticles—A Review. Trop J Pharm Res 5(1):561–573
93. Vauthier C, Bertholon I, Labarre D (2010) Integrated development of glycobiologics: from discovery to applications in the design of nanoparticular drug delivery systems. In: Pharmaceutical sciences encyclopedia: Drug discovery, development, and manufacturing, pp 1–36
94. Tsai IY, Kuo CC, Tomczyk N, Stachelek SJ, Composto RJ, Eckmann DM (2011) Human macrophage adhesion on polysaccharide patterned surfaces. Soft Matter 7(7):3599–3606
95. Patel J, Patel A (2016) Artificial neural networking in controlled drug delivery. In: Puri M, Pathak Y, Kumar Sutariya V, Tipparaju S, Moreno W (ed), Artificial neural network for drug design, delivery and disposition. Academic Press, pp 195–218
96. Yeo Y, Park K (2004) Control Of Encapsulation Efficiency and Initial Burst in Polymeric Microparticle Systems. Archives of Pharmacol Res 27:1–12
97. Huang X, Brazel CS (2001) On the importance and mechanisms of burst release in matrix-controlled drug delivery systems. J Controlled Rele 73:121–136
98. Sheikh Hasan A, Socha M, Lamprecht A, Ghazouani FE, Sapin A, Hoffman M, Maincent P, Ubrich N (2007) Effect of the microencapsulation of nanoparticles on the reduction of burst release. Int J Pharm 344:53–61
99. Hombreiro-Pérez M, Siepmann J, Zinutti C, Lamprecht A, Ubrich N, Hoffman M, Bodmeier R, Maincent P (2003) Non-degradable microparticles containing a hydrophilic and/or a lipophilic drug: preparation, characterization and drug release modelling. J Controlled Release 88:413–428
100. Karasulu E, Yes_ImKarasulu H, Ertan G, Kirilmaz L, Güneri T (2003) Extended release lipophilic indomethacin microspheres: formulation factors and mathematical equations fitted drug release rates. Eur J Pharm Sci 19:99–104
101. Bodmeier R, Chen H (1989) Preparation and characterization of microspheres containing the anti-inflammatory agents, indomethacin, ibuprofen, and ketoprofen. J Controlled Release 10:167–175
102. Wang J, Wang BM, Schwendeman SP (2004) Mechanistic evaluation of the glucose induced reduction in initial burst release of octreotide acetate from Poly (D, L Lactide-Co-Glycolide) microspheres. Biomaterials 25:919–1927

103. Castellanos IJ, Carrasquillo KG, Lopez JD, Alvarez M, Griebenow K (2001) Encapsulation of bovine serum albumin in poly (Lactide-Co-Glycolide) microspheres by the solid-in-oil-in-water technique. J Pharm Pharmacol 53:167–178
104. Thote AJ, Gupta RB (2005) Formation of nanoparticles of a hydrophilic drug using supercritical carbon dioxide and microencapsulation for sustained release. Nanomedicine 1:85–90
105. Thote AJ, Chappell JT, Gupta RB, Kumar R (2005) Reduction in the initial-burst release by surface crosslinking of PLGA microparticles containing hydrophilic or hydrophobic drugs. Drug Dev Ind Pharm 31:43–57
106. Ying XY, Du YZ, Chen WW, Yuan H, Hu FQ (2008) Preparation and characterization of modified lipid nanoparticles for doxorubicin controlled release. Pharmazie 63(12):878–882
107. Chen YC, Ho HO, Liu DZ, Siow WS, Sheu MT (2015) Swelling/floating capability and drug release characterizations of gastroretentive drug delivery system based on a combination of hydroxyethyl cellulose and sodium carboxymethyl cellulose. PLoS ONE 10(1):
108. Williams DH (1986) Definitions in Biomaterials: Proceedings of a Consensus Conference of the European Society for Biomaterials, Chester, England
109. Shin H, Temenoff JS, Mikos AG (2003) *In Vitro* Cytotoxicity of unsaturated oligo [Poly (Ethylene Glycol) Fumarate] Macromers and their cross-linked hydrogels. Biomacromol 4:552–560
110. Hoffman AS (2012) Hydrogels for biomedical applications. Advanced Drug Delivery Reviews, 64(SUPPL.), 18–23
111. Liu J, Zhan X, Wan J, Wang Y, Wang C (2015) Review for carrageenan-based pharmaceutical biomaterials: Favourable physical features versus adverse biological effects. Carbohyd Polym 121:27–36
112. Popa EG, Carvalho PP, Dias AF, Santos TC, Santo VE, Marques AP, Reis RL (2014) Evaluation of the *in vitro* and *in vivo* biocompatibility of carrageenan based hydrogels. J Biomed Mater Res-Part A 102(11):4087–4097

48. Dustgani A, Vasheghani E, Imani M (2008) Preparation of chitosan nanoparticles loaded by dexamethasone sodium phosphate. Ianian J Pharm Sci 4(2):111–114
49. Hassani S, Laouini A, Fessi H, Charcosset C (2015) Preparation of chitosan-TPP nanoparticles using microengineered membranes—effect of parameters and encapsulation of tacrine. Colloids Surf A Physicochem Eng Asp. https://doi.org/10.1016/j.colsurfa.2015.04.006
50. Al-nemrawi NK, Alsharif SSM, Alzoubi KH, Rami Q (2019) Preparation and characterization of insulin chitosan-nanoparticles loaded in buccal films. Pharm Dev Technol 24(8):967–974. https://doi.org/10.1080/10837450.2019.1619183
51. Patel BK, Parikh RH, Aboti PS (2013) Development of oral sustained release Rifampicin loaded chitosan nanoparticles by design of experiment. J. Drug Deliv
52. Farias da Silva Furtado GT, Bizerra Fideles T, Leal Cruz RCA, de Lima Souza JW, Rodriguez Barbero M, Lia Fook MV (2018) Chitosan/NaF particles prepared via ionotropic gelation: evaluation of particles size and morphology. Mater Res 21(4)
53. Cho AR, Chun YG, Kim BK, Park DJ (2014) Preparation of Chitosan—TPP microspheres as resveratrol carriers 79(4):568–576. https://doi.org/10.1111/1750-3841.12395
54. Yuan D, Jacquier JC, Riordan EDO (2017) Entrapment of protein in chitosan-tripolyphosphate beads and its release in an in vitro digestive model. Food Chem 229:495–501. https://doi.org/10.1016/j.foodchem.2017.02.107
55. Gomathi T, Sudha PN, Florence JAK, Venkatesan J, Sukumaran A (2017) Fabrication of letrozole formulation using chitosan nanoparticles through ionic gelation method. Int J Biol Macromol. https://doi.org/10.1016/j.ijbiomac.2017.01.147
56. Yegappan R, Selvaprithiviraj V, Amirthalingam S, Jayakumar R (2018) Carrageenan based hydrogels for drug delivery, tissue engineering and wound healing. Carbohydr Polym 198:385–400. https://doi.org/10.1016/j.carbpol.2018.06.086
57. Necas J, Bartosikova L (2013) Carrageenan: a review 4:187–205
58. Takada A, Kadokawa JI (2015) Fabrication and Characterization of Polysaccharide Ion Gels with Ionic Liquids and Their Further Conversion into Value-Added Sustainable Materials. Biomolecules 5(1):244–262. https://doi.org/10.3390/biom5010244
59. Sedayu BB, Cran MJ, Bigger SW (2019) A Review of Property Enhancement Techniques for Carrageenan-Based Films and Coatings. Carbohydr Polym 216:287–302. https://doi.org/10.1016/j.carbpol.2019.04.021
60. Zia KM, Tabasum S, Nasif M, Sultan N, Aslam N, Noreen A, Zuber M (2017) A Review on Synthesis, Properties and Applications of Natural Polymer Based Carrageenan Blends and Composites. Int J Biol Macromol 96:282–301. https://doi.org/10.1016/j.ijbiomac.2016.11.095
61. Liu J, Zhan X, Wan J, Wang Y, Wang C (2015) Review for Carrageenan-Based Pharmaceutical Biomaterials: Favourable Physical Features versus Adverse Biological Effects. Carbohydr Polym 121:27–36. https://doi.org/10.1016/j.carbpol.2014.11.063
62. Sipahigil O, Dortunc B (2001) Preparation and in vitro evaluation of verapamil HCl and ibuprofen containing carrageenan beads. Int J Pharm 228:119–128. https://doi.org/10.1016/S0378-5173(01)00814-6
63. Hariyadi DM, Hendradi E, Sharon N (2019) Development of carrageenan polymer for encapsulation of ciprofloxacin HCL : in vitro characterization 9(1):89–93. https://doi.org/10.25258/ijddt.9.1.14
64. Soukoulis C, Tsevdou M, Andre CM, Cambier S, Yonekura L, Taoukis PS, Hoffmann L (2017) Modulation of Chemical Stability and in Vitro Bioaccessibility of Beta-Carotene Loaded in Kappa-Carrageenan Oil-in-Gel Emulsions. Food Chem 220:208–218. https://doi.org/10.1016/j.foodchem.2016.09.175
65. Leong KH (2011) Carboxymethylation of Kappa-Carrageenan for Intestinal-Targeted Delivery of Bioactive Macromolecules
66. Wasiak I, Kulikowska A, Janczewska M, Michalak M, Cymerman IA, Nagalski A, Kallinger P, Szymanski WW, Ciach T (2016) Dextran Nanoparticle Synthesis and Properties. PLoS ONE 11(1):1–17. https://doi.org/10.1371/journal.pone.0146237

67. Kutscher M, Sin W, Werner V, Lorenz U, Ohlsen K, Meinel L, Hadinoto K, Germershaus O (2015) Influence of Salt Type and Ionic Strength on Self-Assembly of Dextran Sulfate-Ciprofloxacin Nanoplexes. Int J Pharm 486(1–2):21–29. https://doi.org/10.1016/j.ijpharm.2015.03.022
68. Quadrado RFN, Fajardo AR (2018) Microparticles Based on Carboxymethyl Starch/Chitosan Polyelectrolyte Complex as Vehicles for Drug Delivery Systems. Arab. J. Chem. https://doi.org/10.1016/j.arabjc.2018.04.004
69. Lin WC, Yu DG, Yang MC (2005) PH-Sensitive Polyelectrolyte Complex Gel Microspheres Composed of Chitosan/Sodium Tripolyphosphate/Dextran Sulfate: Swelling Kinetics and Drug Delivery Properties. Colloids Surfaces B Biointerfaces 44(2–3):143–151. https://doi.org/10.1016/j.colsurfb.2005.06.010
70. Gatti THH, Eloy JO, Ferreira LMB, Da Silva IC, Pavan FR, Gremião MPD, Chorilli M (2018) Insulin-Loaded Polymeric Mucoadhesive Nanoparticles: Development, Characterization and Cytotoxicity Evaluation. Brazilian J. Pharm. Sci. 54(1):1–10. https://doi.org/10.1590/s2175-97902018000117314
71. Chavan C, Bala P, Pal K, Kale SN (2017) Cross-Linked Chitosan-Dextran Sulphate Vehicle System for Controlled Release of Ciprofloxaxin Drug: An Ophthalmic Application. OpenNano 2:28–36. https://doi.org/10.1016/j.onano.2017.04.002
72. Unagolla JM, Jayasuriya AC (2018) Drug Transport Mechanisms and in Vitro Release Kinetics of Vancomycin Encapsulated Chitosan-Alginate Polyelectrolyte Microparticles as a Controlled Drug Delivery System. Eur J Pharm Sci 114:199–209. https://doi.org/10.1016/j.ejps.2017.12.012
73. Akolade JO, Oloyede HOB, Onyenekwe PC (2017) Encapsulation in Chitosan-Based Polyelectrolyte Complexes Enhances Antidiabetic Activity of Curcumin. J. Funct. Foods 35:584–594. https://doi.org/10.1016/j.jff.2017.06.023
74. Xu Y, Zhan C, Fan L, Wang L, Zheng H (2007) Preparation of Dual Crosslinked Alginate-Chitosan Blend Gel Beads and in Vitro Controlled Release in Oral Site-Specific Drug Delivery System. Int J Pharm 336(2):329–337. https://doi.org/10.1016/j.ijpharm.2006.12.019
75. Mcintosh M, Stone BA, Stanisich VA (2005) Curdlan and Other Bacterial (1 → 3)- β -. Appl Microbiol Biotechnol 1:163–173. https://doi.org/10.1007/s00253-005-1959-5
76. Yan JK, Wang YY, Qiu WY, Wu JY (2017) Construction and Characterization of Nano-sized Curdlan Sulfate/Chitosan Polyelectrolyte Complex toward Drug Release of Zidovudine. Carbohydr Polym 174:209–216. https://doi.org/10.1016/j.carbpol.2017.06.082
77. Chen L, Zheng Y, Feng L, Liu Z, Guo R, Zhang Y (2019) Novel Hyaluronic Acid Coated Hydrophobically Modified Chitosan Polyelectrolyte Complex for the Delivery of Doxorubicin. Int J Biol Macromol 126:254–261. https://doi.org/10.1016/j.ijbiomac.2018.12.215
78. Paini M, Aliakbarian B, Casazza AA, Perego P, Ruggiero C, Pastorino L (2015) Chitosan/Dextran Multilayer Microcapsules for Polyphenol Co-Delivery. Mater Sci Eng, C 46:374–380
79. Anirudhan T, Sekhar C, Vijayamma A (2017) Layer-by-Layer Assembly of Hyaluronic Acid/Carboxymethylchitosan Polyelectrolytes on the Surface of Aminated Mesoporous Silica for the Oral Delivery of 5-Fluorouracil. Eur Polym J. https://doi.org/10.1016/j.eurpolymj.2017.06.033
80. Zhang Y, Chi C, Huang X, Zou Q, Li X, Chen L (2017) Starch-Based Nanocapsules Fabricated through Layer-by-Layer Assembly for Oral Delivery of Protein to Lower Gastrointestinal Tract Highlights. Carbohydr Polym. https://doi.org/10.1016/j.carbpol.2017.04.090
81. Rochin-Wong S, Rosas-Durazo A, Zavala-Rivera P, Maldonado A, Martínez-Barbosa ME, Vélaz I, Tánori J (2018) Drug Release Properties of Diflunisal from Layer-By- Nanocapsules: Effect of Deposited Layers. Polymers (Basel) 10(760):1–16. https://doi.org/10.3390/polym10070760
82. Jeon S, Yoo CY, Park SN (2015) Colloids and Surfaces B: Biointerfaces Improved Stability and Skin Permeability of Sodium Hyaluronate-Chitosan Multilayered Liposomes by Layer-by-Layer Electrostatic Deposition for Quercetin Delivery. Colloids Surfaces B Biointerfaces 129:7–14. https://doi.org/10.1016/j.colsurfb.2015.03.018